MATH PRACTICE WORKBOOK

400+ QUESTIONS YOU NEED TO KILL IN 2nd GRADE

+7 DAYS ONLINE ACCESS TO PREMIUM CONTENT QUIZZES & DRILLS WITH VIDEO EXPLANATIONS

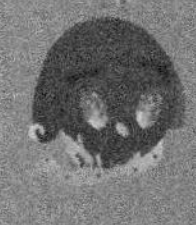

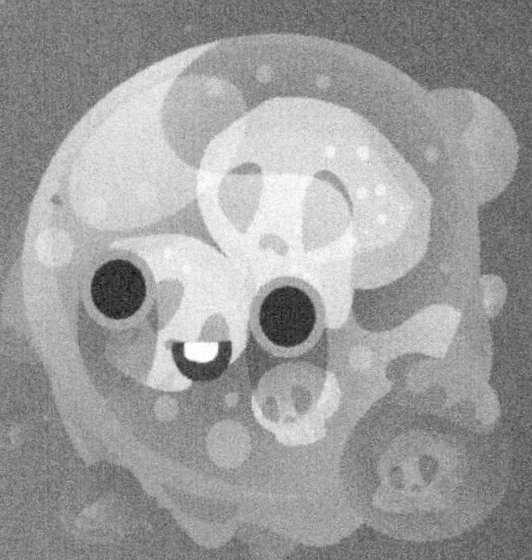

CREKS

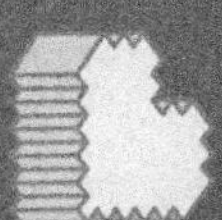

made by **Brain Hunter Prep** with love from **New York**

Want an amazing offer from us?

7 DAY ACCESS

to our online premium content at www.argoprep.com

Online premium content includes practice quizzes and drills with video explanations and an automatic grading system.

Chat with us live at **www.argoprep.com** for this exclusive offer.

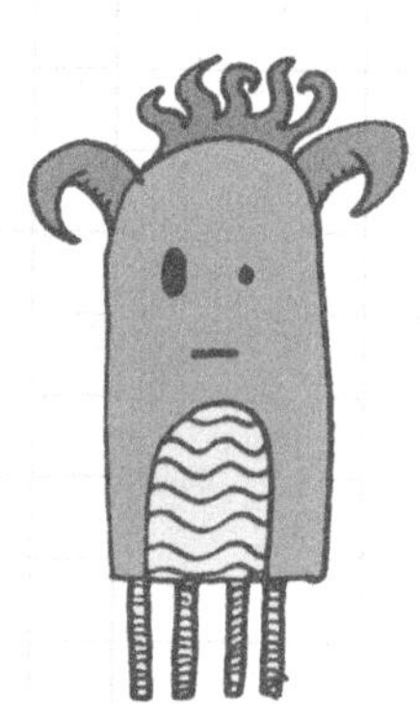

We take great pride in providing our students, parents and educators with the best educational products and customer service. We kindly ask for one minute of your time to leave us an honest review on Amazon for this workbook.

Reviews help us grow our brand, improve our products, and allow customers to learn more about our workbooks. Thanks for letting us be a part of your child's educational journey.

ISBN-13: 9781951048297
Published by Brain Hunter Prep Inc.

Books made by **Brain Hunter Prep** with love from **New York**

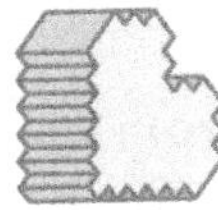

KILL IT SERIES

KILL IT INDIVIDUAL SERIES

Check out our ArgoPrep brand workbooks!

All of our workbooks come equipped with detailed video explanations to make your learning experience a breeze! Visit us at www.argoprep.com

COMMON CORE SERIES

SPECIALIZED HIGH SCHOOL ADMISSIONS TEST

HIGHER LEVEL EXAMS

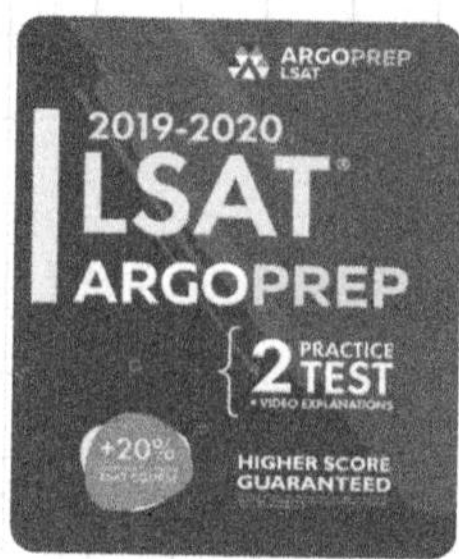

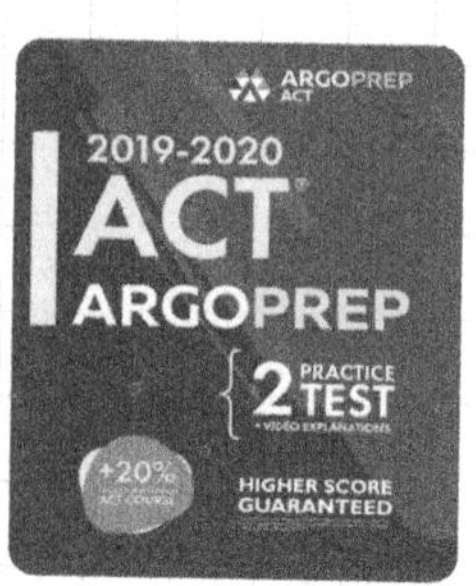

INTRODUCING MATH!

Introducing Math! by ArgoPrep is an award-winning series created by certified teachers to provide students with high-quality practice problems. These workbooks include topic overviews with instruction, practice questions, answer explanations along with digital access to video explanations. Practice in confidence - with ArgoPrep!

YOGA MINDFULNESS

If you are looking for a fun way to engage with your children while helping them build a mindful, engaged and healthy lifestyle, Frogyogi's Yoga Stories for Kids and Parents is the perfect book for you and your family!

KIDS SUMMER ACADEMY SERIES

ArgoPrep's **Kids Summer Academy** series helps prevent summer learning loss and gets students ready for their new school year by reinforcing core foundations in math, english and science. Our workbooks also introduce new concepts so students can get a head start and be on top of their game for the new school year!

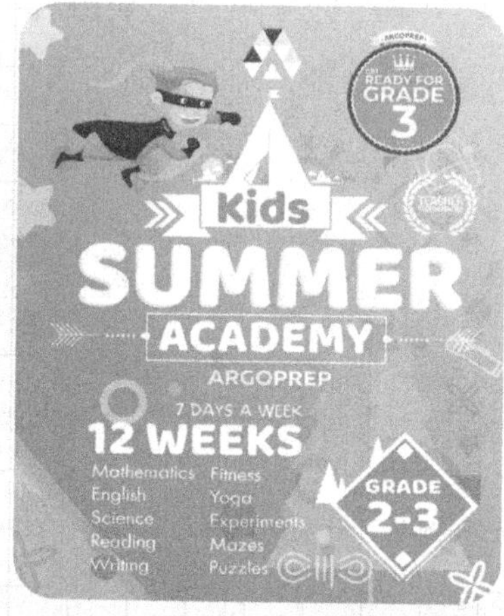

Welcome! Meet the ArgoPrep heroes.

Are you ready to go on an incredible adventure and complete your journey with them to become a **SUPER** student?

WATER FIRE

MYSTICAL NINJA

FIRESTORM WARRIOR

RAPID NINJA

CAPTAIN ARGO

THUNDER WARRIOR

ADRASTOS THE SUPER WARRIOR

Our **Kids Summer Academy** series by **ArgoPrep** is designed to keep students engaged with fun graphics and activities. Our curriculum is aligned with state standards to help your child prepare for their new school year.

GRADE 2 MATH PRACTICE WORKBOOK

TABLE OF CONTENTS

CHAPTER 1

GRADE 2

1. What addition equation represents joining these two groups?

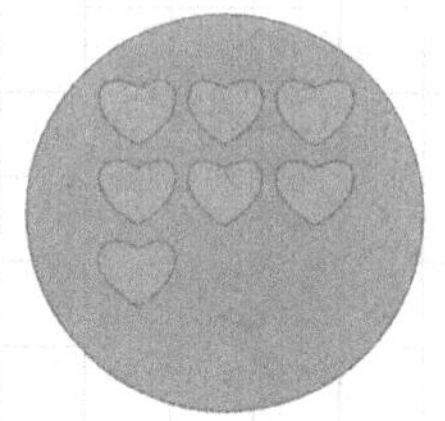

A. 10 + 6 = 16
B. 10 + 7 = 17
C. 9 + 7 = 16
D. 10 + 7 = 18

2. What addition equation represents joining these two groups?

A. 15 + 12 = 27
B. 12 + 15 = 25
C. 14 + 11 = 25
D. 15 + 12 = 26

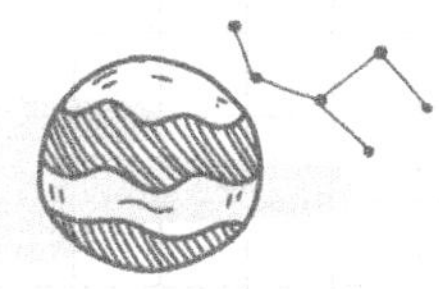

3. Write an addition equation to match this picture.

4. Draw a picture to represent the following addition equation.

12 + 9 = 21

5. Draw a picture to represent the following addition equation.

19 + 6 = 25

6. What addition equation matches the picture?

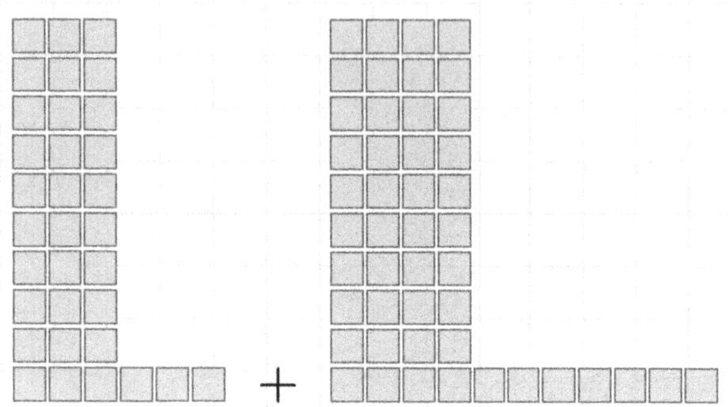

A. 22 + 57 = 79
B. 33 + 45 = 78
C. 23 + 37 = 60
D. 33 + 47 = 80

TIP: Don't count all the blocks individually! Instead, count the groups of tens and then count the ones.

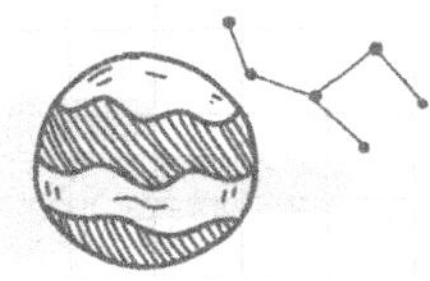

7. What addition equation matches the picture?

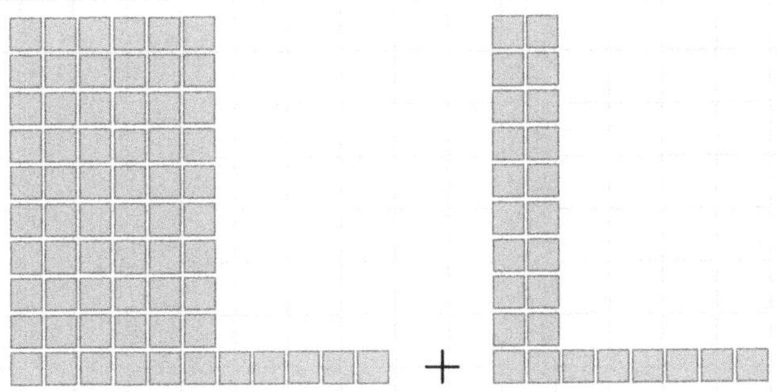

A. 55 + 25 = 80
B. 74 + 25 = 99
C. 65 + 26 = 91
D. 70 + 26 = 96

8. What addition equation represents joining these three groups?

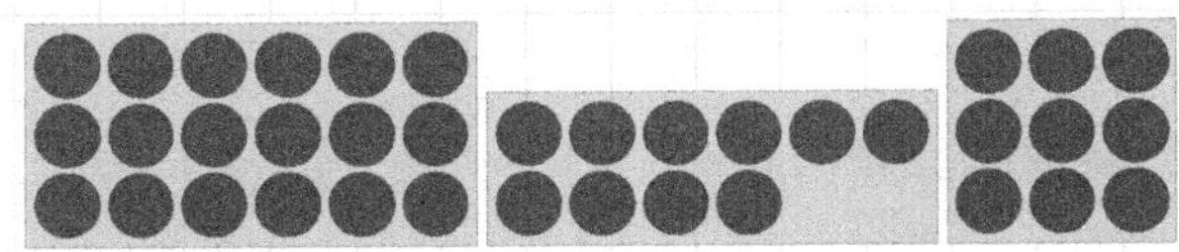

A. 15 + 9 + 9 = 33
B. 20 + 10 + 8 = 28
C. 17 + 9 + 8 = 34
D. 18 + 10 + 9 = 37

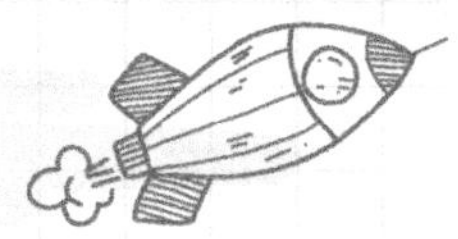

9. What addition equation represents joining these three groups?

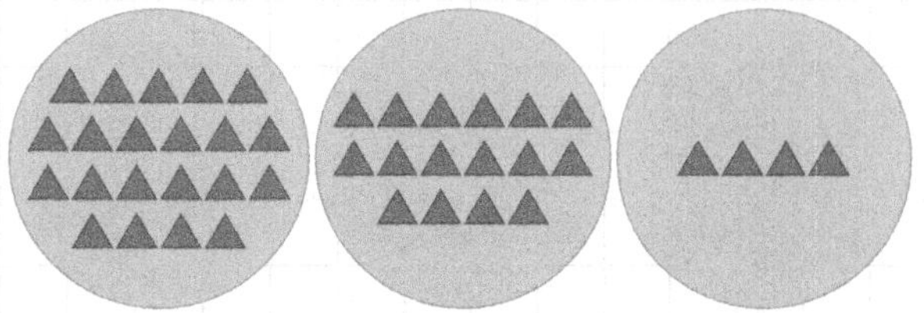

A. 20 + 15 + 4 = 39
B. 21 + 16 + 5 = 42
C. 21 + 16 + 4 = 41
D. 19 + 14 + 3 = 36

10. Draw a picture to represent the following addition equation.

17 + 15 + 2 = 34

11. Draw a picture to represent the following addition equation.

31 + 9 + 10 = 50

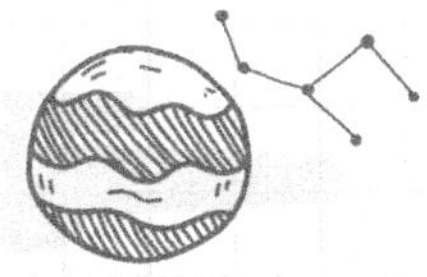

12. Draw a picture to find the missing number.

42 + ? = 65

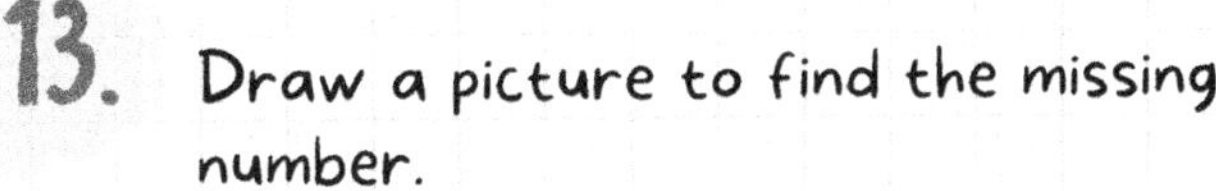

13. Draw a picture to find the missing number.

50 + ? = 70

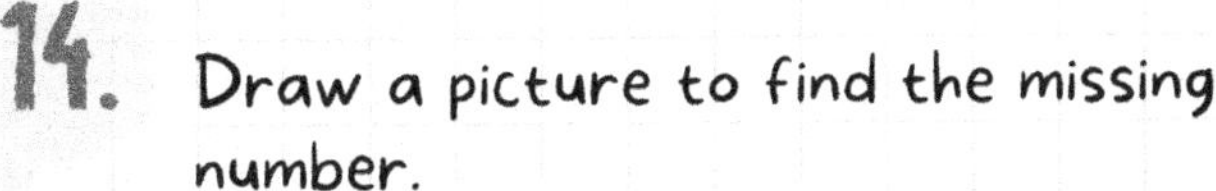

14. Draw a picture to find the missing number.

? + 25 = 50

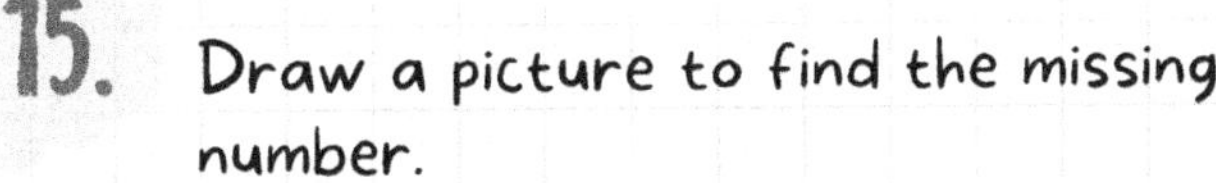

15. Draw a picture to find the missing number.

? + 30 = 85

16. Ava has 35 stickers. When she goes to the store with her mom, she gets a large pack of 62 stickers. When she gets home, she put them all on a poster board. How many stickers are on the poster board? Show your work!

17. Jake collected 47 rocks in his backyard. His sister collected some too while she was at school. They put them together in their rock collection and had 100 rocks. How many rocks did Jake's sister collect? Show your work!

TIP: When reading word problems, try to visualize what is happening. This will help you figure out if you need to add or subtract!

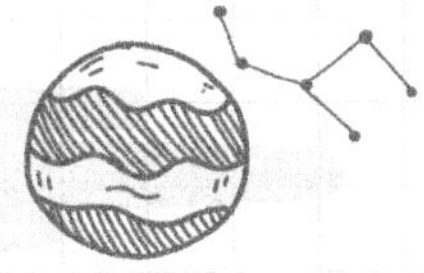

18. Levi counted the money in his piggy bank and had 32 cents. He wanted to buy a candy bar from the snack stand that cost 50 cents. He asked his mom for the rest of the money to buy the candy bar, and she gave it to him. How much money did Levi's mom give to him?

19. Gina went to the blueberry farm and picked 43 blueberries. Her sister picked 42 blueberries. Then, they combined them all in a bowl to share for a snack. Gina ate 10 right away. How many blueberries are in the bowl now? Show your work!

20. Abby was going to have a lemonade stand. She bought 60 cups at the store. When she got home she found a pack of 33 cups that she had leftover from her sale last year. At the lemonade sale, she sold 73 cups of lemonade! How many cups did she have left?

21. What subtraction equation represents this picture?

A. 20 - 10 = 10
B. 22 - 11 = 11
C. 11 - 11 = 0
D. 23 - 12 = 11

22. What subtraction equation represents this picture?

A. 32 - 14 = 18
B. 18 - 14 = 4
C. 30 - 15 = 15
D. 32 - 15 = 17

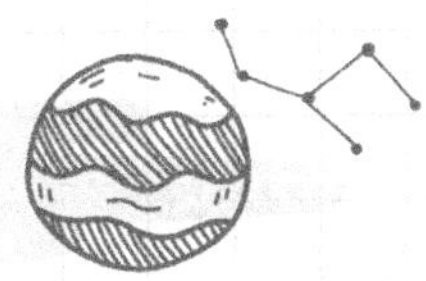

23. Write a subtraction equation to match this picture.

24. Draw a picture to represent the following subtraction equation.

$$45 - 22 = 23$$

25. Draw a picture to represent the following subtraction equation.

$$31 - 20 = 11$$

26. What subtraction equation matches this picture?

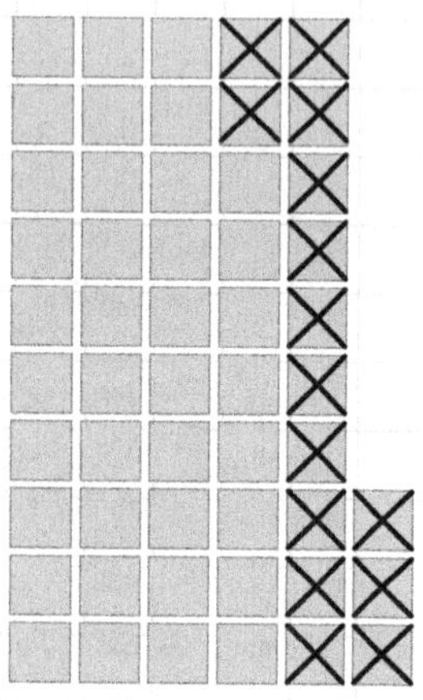

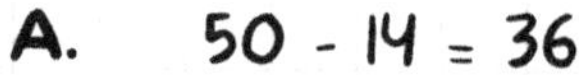

A. 50 - 14 = 36
B. 60 - 15 = 45
C. 50 - 3 = 47
D. 53 - 15 = 38

TIP: When writing subtraction equations, remember to start with the whole group and then subtract the number that is crossed out. You are left with the difference.

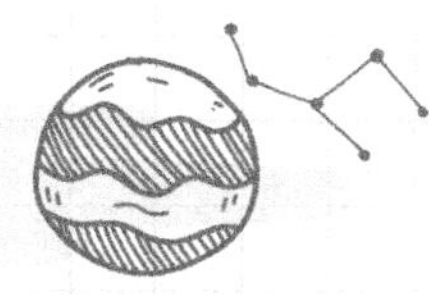

27. What subtraction equation matches this picture?

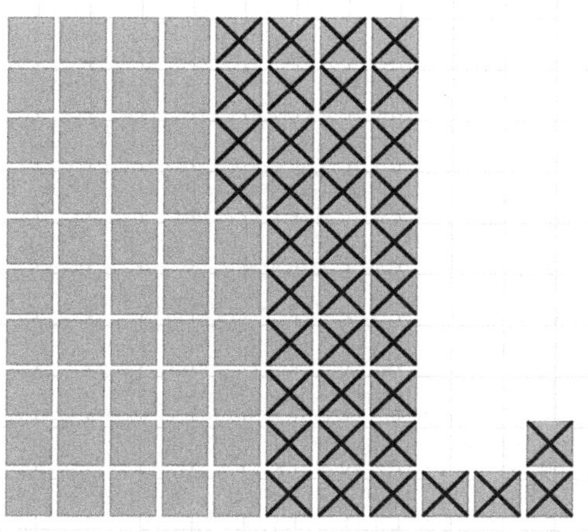

A. 84 - 48 = 36
B. 89 - 39 = 50
C. 84 - 38 = 46
D. 74 - 46 = 28

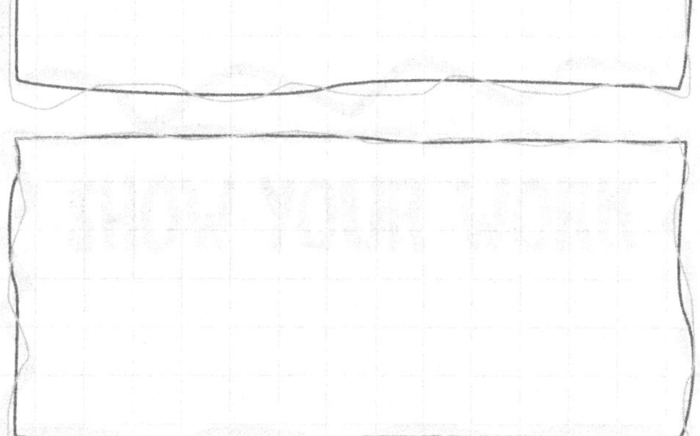

28. Draw a picture to find the missing number.

54 - ? = 37

29. Draw a picture to find the missing number.

95 - ? = 21

30. Draw a picture to find the missing number.

? - 28 = 49

31. Draw a picture to find the missing number.

? - 18 = 44

32. 20 dogs were for sale at the pet shop. If 5 were sold, how many were left? Find the subtraction equation that shows the difference.

A. 15 - 5 = 10
B. 5 - 20 = 5
C. 20 - 5 = 15
D. 20 - 15 = 5

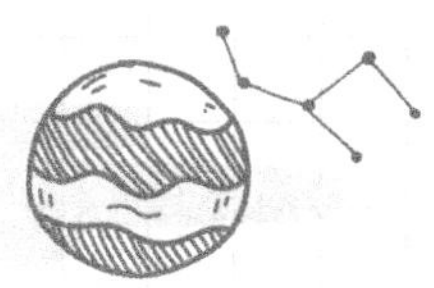

33. The bakery made **32** cookies. Mrs. Smith came in and asked to buy **24** cookies for her daughter's birthday party. How many cookies were left at the bakery?

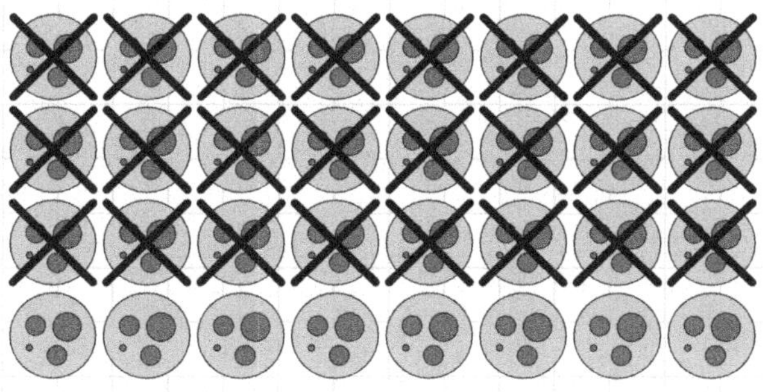

A. 32 - 24 = 8
B. 30 - 25 = 5
C. 24 - 8 = 16
D. 32 - 21 = 11

34. There were **25** fish swimming in the pond. A fisherman went out fishing for the day and caught a few fish. There were **20** fish left in the pond that night. How many fish did the fisherman catch? Show your work!

35. Sophia collected **32** seashells at the beach! When she got home, she painted them for her family and gave some away to her sisters. She had **10** left for her cousins. How many shells did Sophia give to her sisters? Show your work!

36. My mom planted a beautiful flower garden with red flowers. She said I could pick a few for Grandma. I picked **15** flowers. There were **56** flowers left in the garden. How many flowers did my mom plant? Show your work!

TIP: Related addition or subtraction facts can help you solve word problems. You can solve the problems in more than one way!

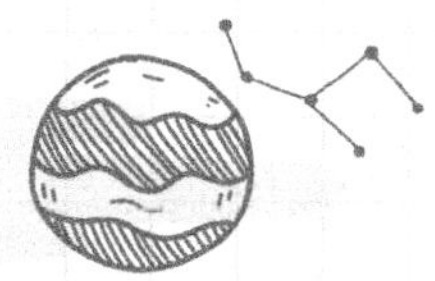

37. Andrew was outside bug collecting and caught 41 bugs. When he opened the jar to show his mom, 12 bugs flew out. He was able to catch 5 of them before they got away. How many bugs did he have left in the jar? Show your work!

38. I had a big box of 64 crayons. My friend, Brooke, had a box of 50 crayons. How many more crayons did I have? Show your work!

39. Natalie is 45 inches tall, and her brother is 36 inches tall. How much taller is Natalie than her brother? Show your work!

40. Cody is a big dog. He weighs 65 pounds. Scout is a small dog. She weighs 12 pounds. How much more does Cody weigh than Scout? Show your work!

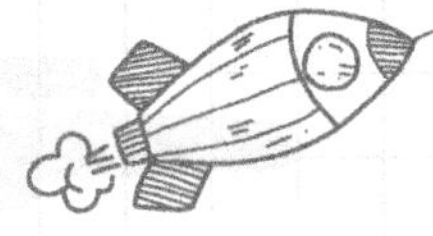

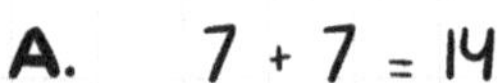

41. Which double fact would help you solve the following problem? 7 + 8 = 15

A. 7 + 7 = 14
B. 6 + 6 = 12
C. 10 + 10 = 20
D. 5 + 5 = 10

42. Which double fact would help you solve the following problem? 10 + 9 = 19

A. 8 + 8 = 16
B. 7 + 7 = 14
C. 9 + 9 = 18
D. 4 + 4 = 8

43. Which double fact would help you solve the following problem? 6 + 7 = 13

A. 5 + 5 = 10
B. 3 + 3 = 6
C. 10 + 10 = 20
D. 7 + 7 = 14

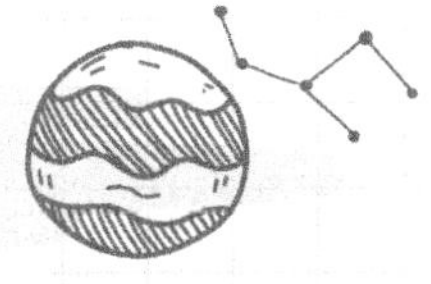

44. Which double fact would help you solve the following problem?
4 + 5 = 9

A. 2 + 2 = 4
B. 5 + 5 = 10
C. 6 + 6 = 12
D. 7 + 7 = 14

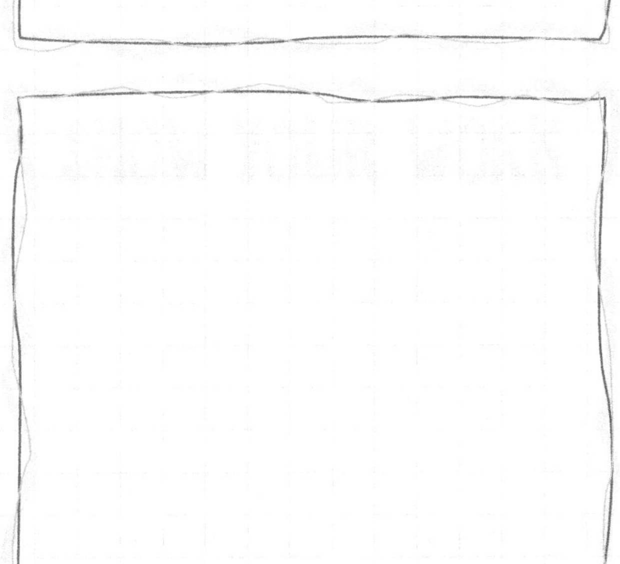

45. Here is the double fact 6 + 6 = 12. Write an equation that would would be a near double.

46. Here is the double fact 8 + 8 = 16. Write an equation that would would be a near double.

47. Write a double fact addition equation.

48. Write a double fact addition equation.

49. Write a double fact subtraction equation.

50. Write a double fact subtraction equation.

TIP: In a subtraction double fact, the difference will be zero!

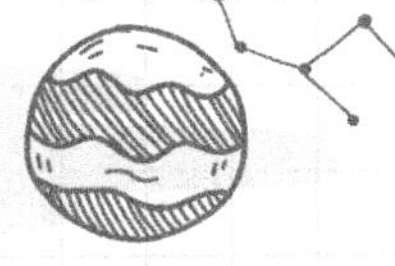

51. If using the strategy to make 10, which problem would this 10 frame help you solve?

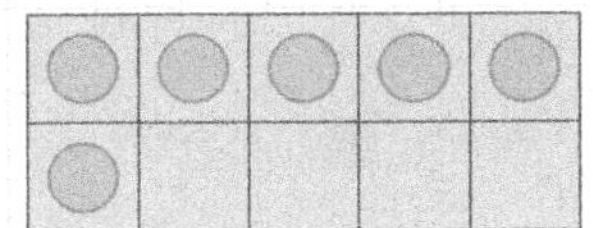

A. 9 + 7 = 16
B. 10 + 7 = 17
C. 8 + 8 = 16
D. 10 + 8 = 18

52. If using the strategy to make 10, which problem would this 10 frame help you solve?

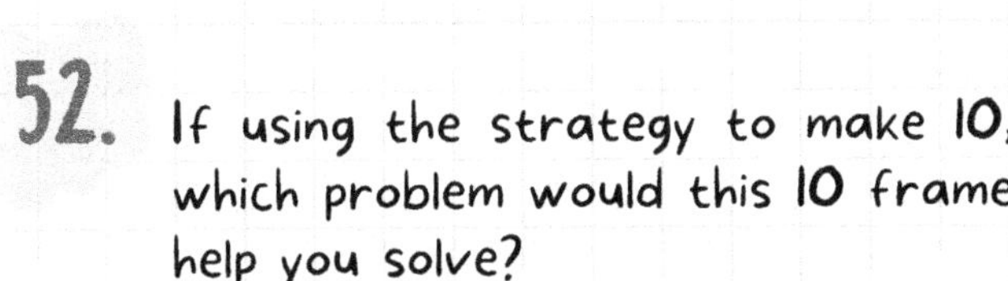

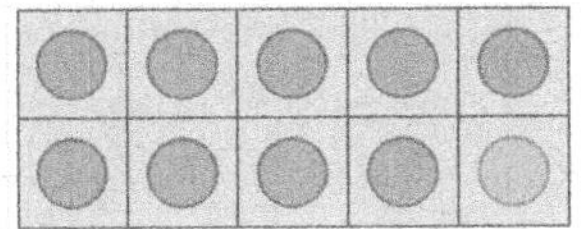

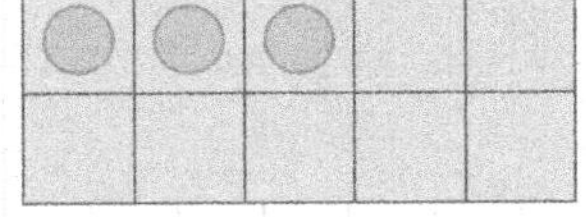

A. 10 + 4 = 14
B. 9 + 5 = 14
C. 9 + 4 = 13
D. 10 + 2 = 12

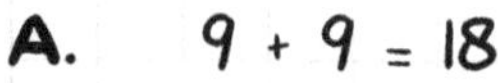

53. If you know that $10 + 7 = 17$, then it should help you solve which problem?

A. $9 + 9 = 18$
B. $10 + 5 = 15$
C. $9 + 8 = 17$
D. $10 + 9 = 19$

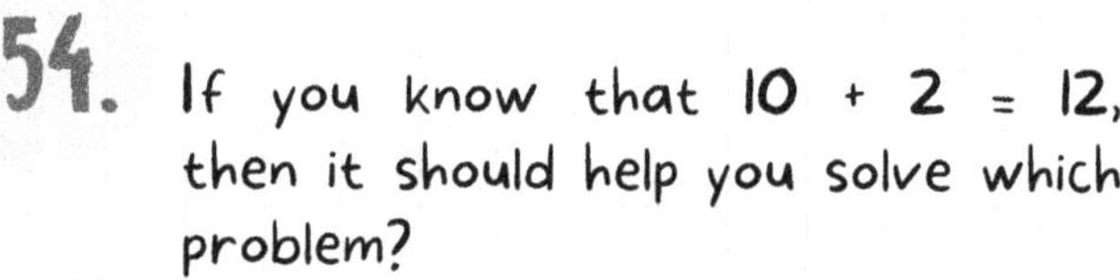

54. If you know that $10 + 2 = 12$, then it should help you solve which problem?

A. $6 + 6 = 12$
B. $10 + 4 = 14$
C. $9 + 4 = 13$
D. $9 + 3 = 12$

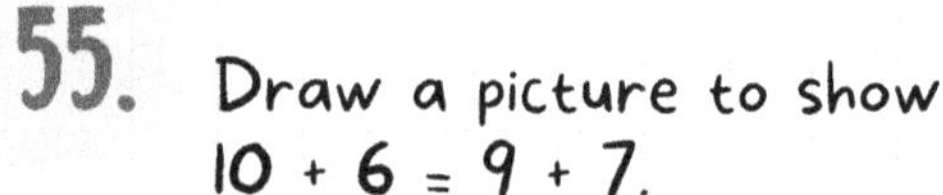

55. Draw a picture to show $10 + 6 = 9 + 7$.

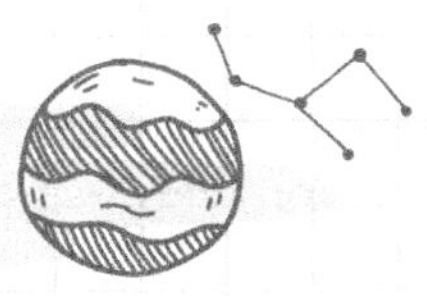

56. Draw a picture to show 9 + 6 = 10 + 5.

57. Which addition fact would help you solve the subtraction problem 16 - 5 = 11?

A. 11 + 5 = 16
B. 10 + 1 = 11
C. 16 + 5 = 21
D. 5 + 1 = 6

58. Which addition fact would help you solve the subtraction problem 19 - 3 = 16?

A. 8 + 8 = 16
B. 10 + 6 = 16
C. 3 + 16 = 19
D. 10 + 9 = 19

OPERATIONS AND ALGEBRAIC THINKING

SECTION 2: ADD AND SUBTRACT WITHIN 20 USING MENTAL STRATEGIES

59. Which addition fact would help you solve the subtraction problem 20 - 7 = 13?

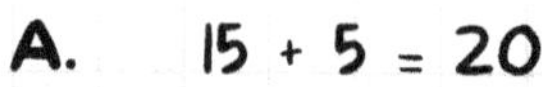

A. 15 + 5 = 20
B. 10 + 10 = 20
C. 13 + 7 = 20
D. 10 + 3 = 13

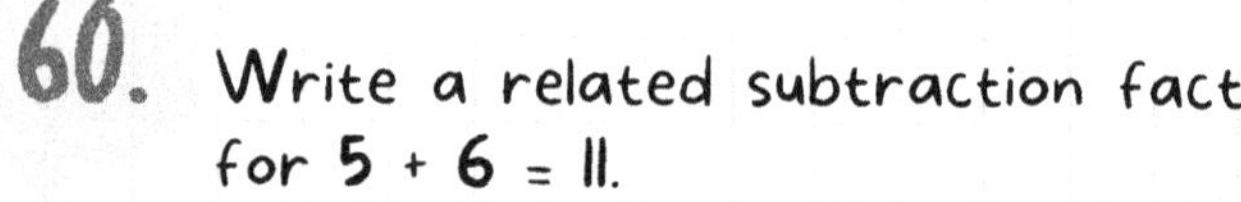

60. Write a related subtraction fact for 5 + 6 = 11.

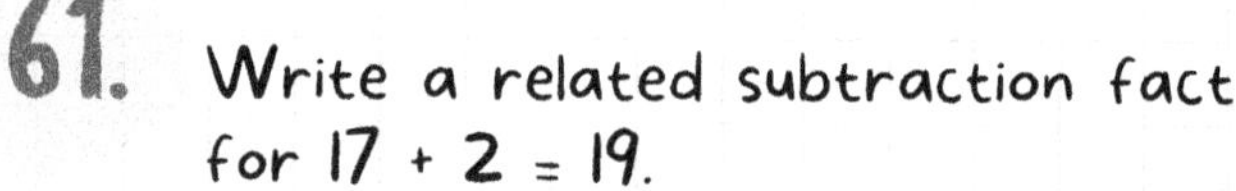

61. Write a related subtraction fact for 17 + 2 = 19.

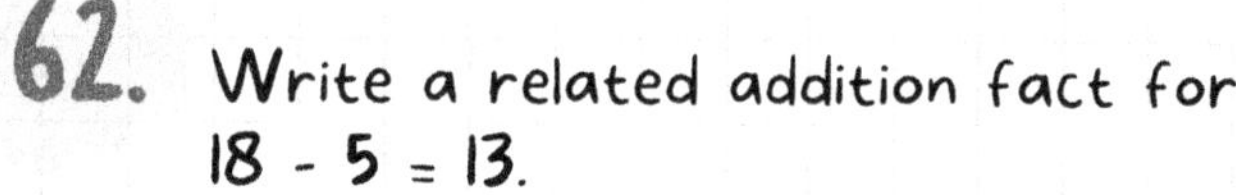

62. Write a related addition fact for 18 - 5 = 13.

TIP: There are four related facts in a fact family! There are two addition and two subtraction facts. For example: 3+1=4, 1+3=4, 4-3=1, 4-1=3.

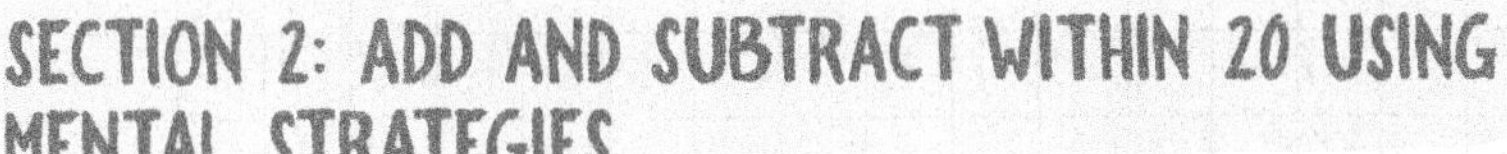

63. Complete these addition problems.

18 + 0 =	4 + 0 =	0 + 19 =
7 + 0 =	0 + 20 =	0 + 15 =

64. Complete these subtraction problems.

12 - 0 =	0 - 0 =	17 - 0 =
6 - 0 =	10 - 0 =	14 - 0 =

65. Solve these addition problems.

4 + 4 =	10 + 10 =	7 + 7 =
6 + 6 =	9 + 9 =	3 + 3 =

66. Complete these subtraction problems.

12 - 12 =	2 - 2 =	5 - 5 =
1 - 1 =	8 - 8 =	7 - 7 =

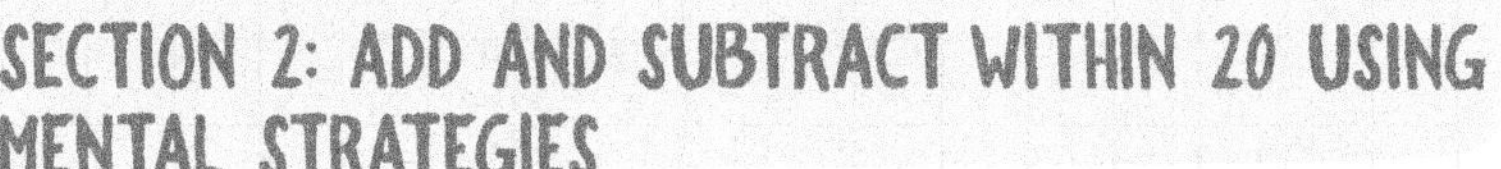

67. Solve these addition problems as quickly as possible.

17 + 3 =	16 + 2 =	9 + 10 =
10 + 4 =	5 + 15 =	7 + 6 =
14 + 4 =	1 + 12 =	4 + 16 =

68. Solve these subtraction problems as quickly as possible.

16 - 5 =	17 - 2 =	13 - 10 =
12 - 4 =	9 - 8 =	14 - 6 =
18 - 7 =	19 - 12 =	17 - 13 =

69. Solve these mixed addition and subtraction problems as quickly as possible.

8 + 12 =	16 + 2 =	20 - 10 =
3 + 15 =	8 + 8 =	19 - 7 =
17 - 7 =	15 - 10 =	2 + 13 =

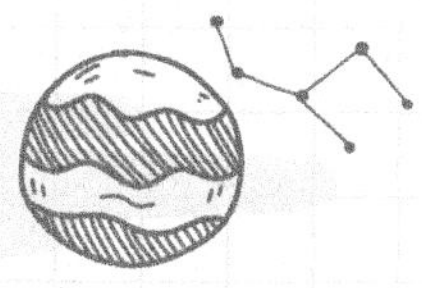

70. What is the missing number in this addition equation? 16 = ? + 10

A. 7
B. 6
C. 26
D. 10

SHOW YOUR WORK

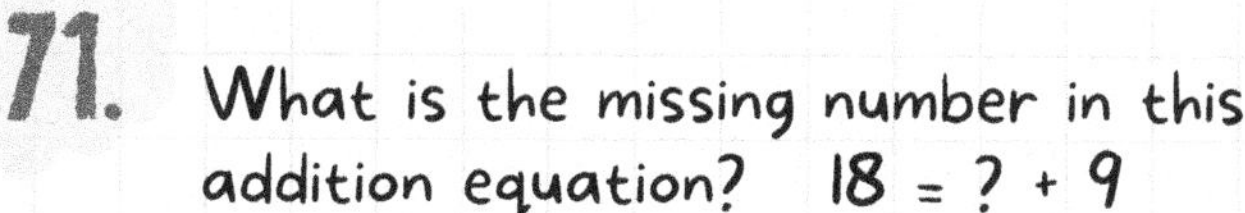

71. What is the missing number in this addition equation? 18 = ? + 9

A. 27
B. 10
C. 9
D. 8

72. What is the missing number in this addition equation? 15 = 8 + ?

A. 9
B. 11
C. 25
D. 7

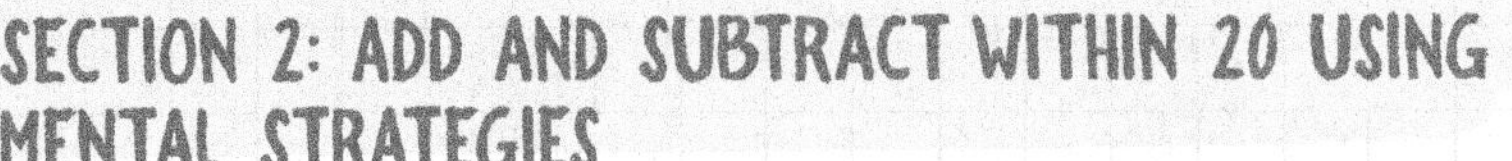

73. What is the missing number in this addition equation? 12 = ? + 0

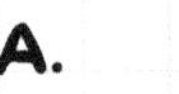

A. 1
B. 4
C. 12
D. 0

74. What is the missing number in this subtraction equation? 8 - ? = 1

A. 7
B. 9
C. 6
D. 10

TIP: To find a missing number in an equation, you can start at the known number and count on until you get to the answer or you can use a related fact.

75. What is the missing number in this subtraction equation? 12 - ? = 6

A. 18
B. 7
C. 6
D. 5

76. What is the missing number in this subtraction equation? ? - 10 = 8

A. 18
B. 9
C. 1
D. 20

77. What is the missing number in this subtraction equation? ? - 7 = 6

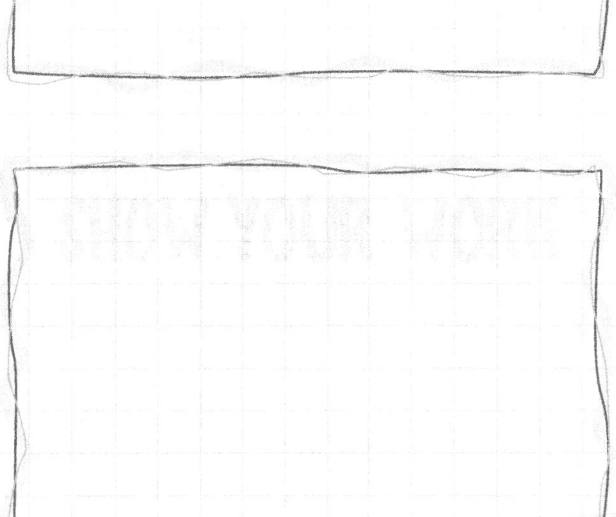

A. 1
B. 14
C. 15
D. 13

78. Solve these addition equations with a missing addend.

15 + __ = 20	__ + 13 = 18	5 + __ = 11
__ + 8 = 10	7 + __ = 14	__ + 0 = 17

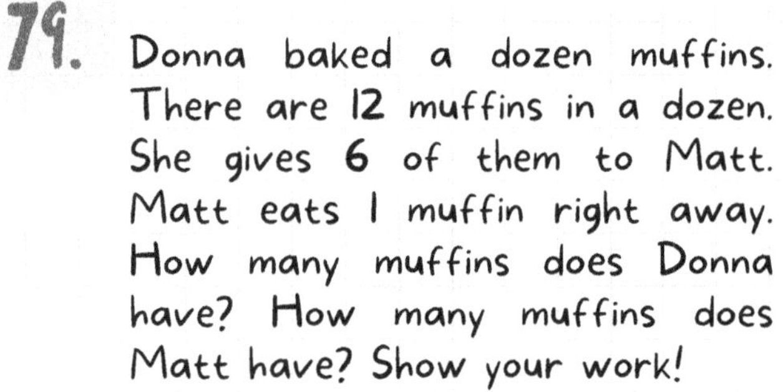

79. Donna baked a dozen muffins. There are 12 muffins in a dozen. She gives 6 of them to Matt. Matt eats 1 muffin right away. How many muffins does Donna have? How many muffins does Matt have? Show your work!

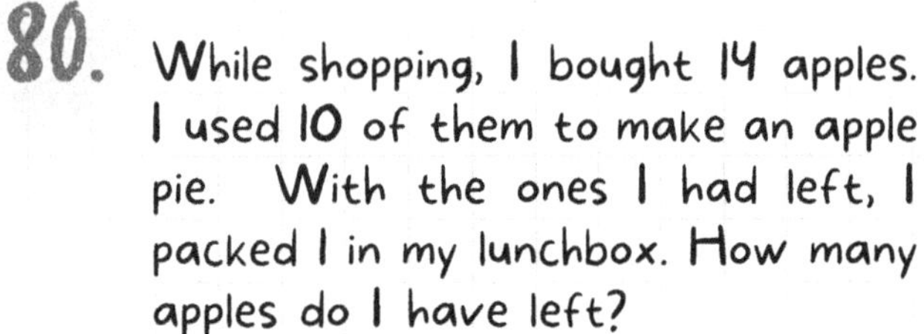

80. While shopping, I bought 14 apples. I used 10 of them to make an apple pie. With the ones I had left, I packed 1 in my lunchbox. How many apples do I have left?

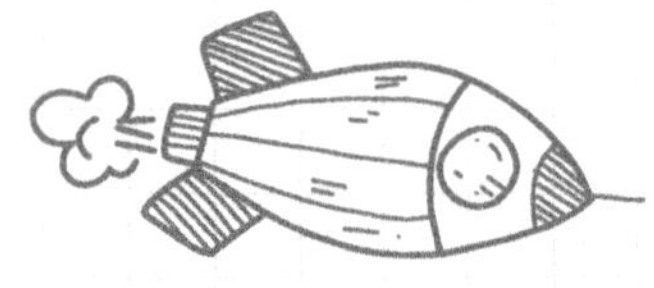

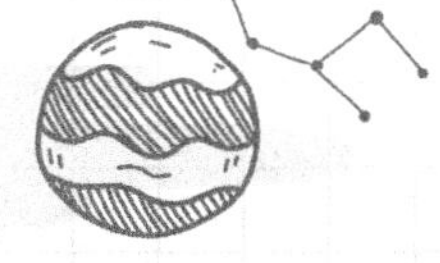

81. How many apples are there? Circle odd or even.

_____ apples Odd Even

SHOW YOUR WORK

82. How many watermelons are there? Circle odd or even.

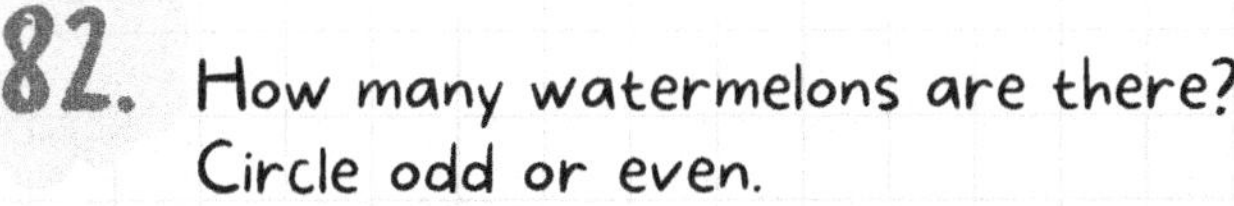

_____ watermelon Odd Even

SHOW YOUR WORK

83. How many bananas are there? Circle odd or even.

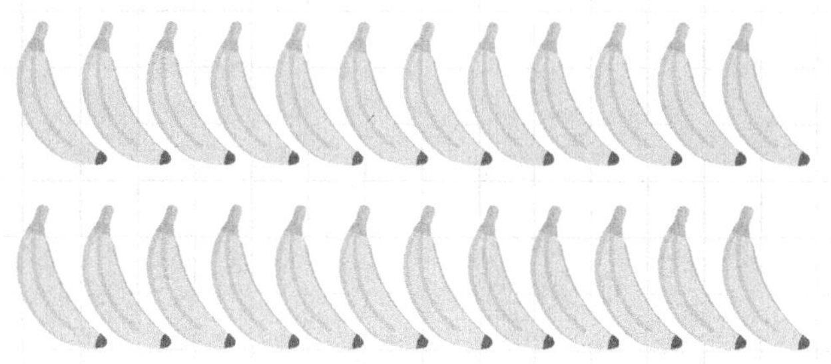

______ bananas Odd Even

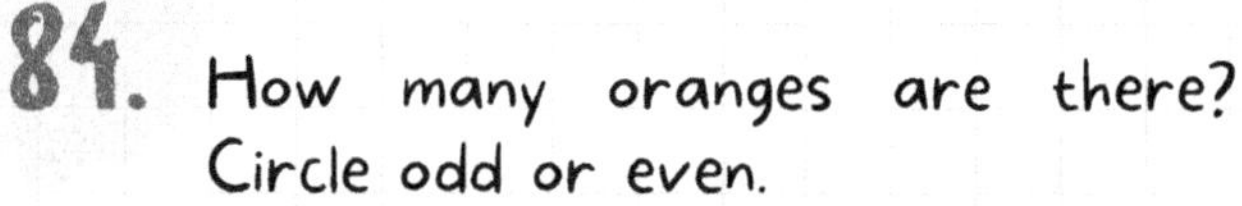

84. How many oranges are there? Circle odd or even.

______ oranges Odd Even

TIP: When you count by 2s, you say all of the even numbers!

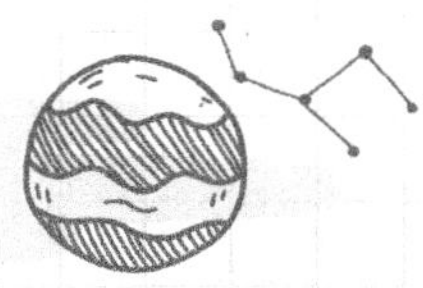

85. Pick a single digit odd number and draw an illustration to show why it is odd.

86. Pick a single digit even number and draw an illustration to show why it is even.

87. Pick a double digit odd number and draw an illustration to show why it is odd.

88. Pick a double digit even number and draw an illustration to show why it is even.

89. Write an addition equation with 2 equal addends to match the picture.

90. Write an addition equation with 2 equal addends to match the picture.

91. Which picture shows 4 + 4 = 8?

A.

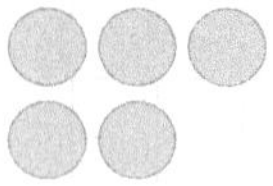

B.

C.

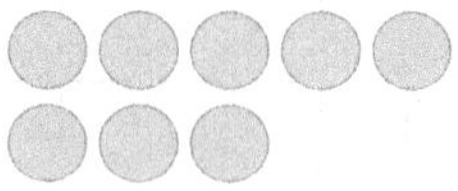

D.

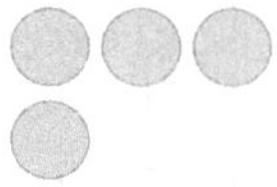

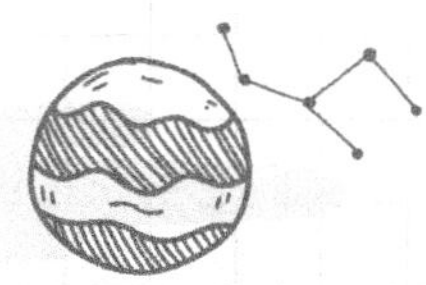

92. Which picture shows 7 + 7 = 14?

A. 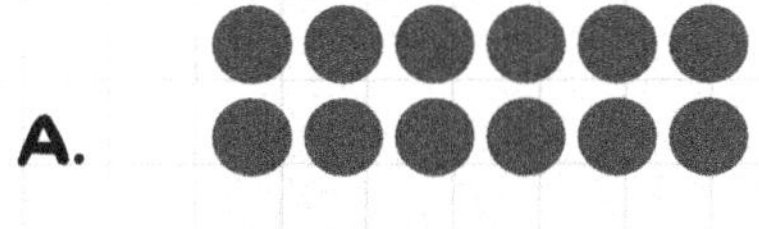

B.

C. 

D.

93. I have **5** people in my family. How many eyes do we have in all? Draw a picture to show your work!

94. There are **6** dogs. How many feet in all? Draw a picture to show your work!

95. There are **8** people and **15** sneakers. Are there enough sneakers for everyone? Draw a picture to show your work!

96. There are 10 children and **20** mittens. Are there enough mittens for everyone? Draw a picture to show your work.

97. Count by **2**'s. Write the number below the picture.

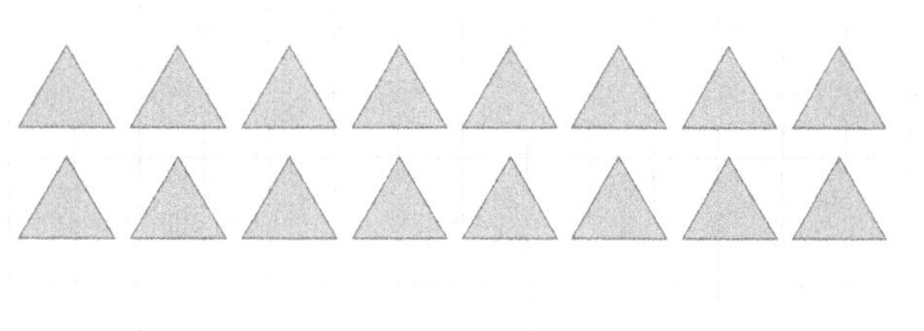

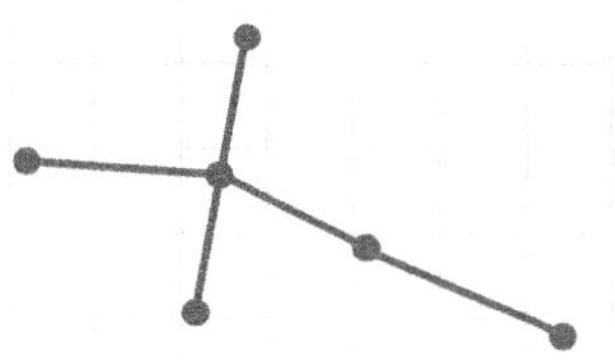

TIP: The sum of double facts are always even numbers. Memorizing your double facts will help you know if a number is even or odd!

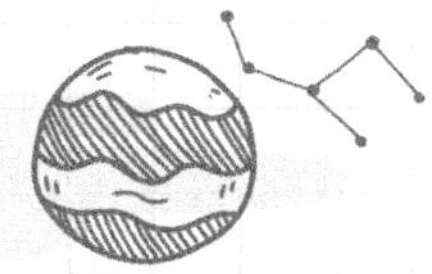

98. Count by 2's. Write the number below the picture.

__ __ __ __ __ __ __ __ __ __ __

99. What even number is greater than 9 and less than 11?

100. What odd number is less than 16 and greater than 14?

101. Write an addition sentence with equal addends to match this array:

102. Write an addition sentence with equal addends to match this array:

103. Write an addition sentence with equal addends to match this array:

104. Write an addition sentence with equal addends to match this array:

105. Write an addition sentence with equal addends to match this array:

106. Write an addition sentence with equal addends to match this array:

107. Write an addition sentence with equal addends to match this array:

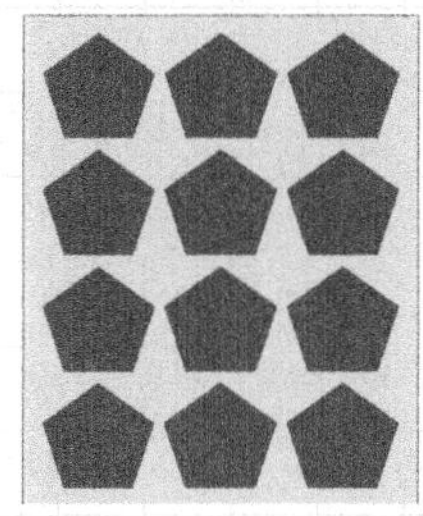

108. Write an addition sentence with equal addends to match this array:

109. Write an addition sentence with equal addends to match this array:

TIP: To write an addition equation to match an array, count how many are in each row. That will be your addend. Then count how many rows to see how many addends you need to get the sum.

110. Write an addition sentence with equal addends to match this array:

111. Draw an array to match the addition sentence $4 + 4 + 4 = 12$.

112. Draw an array to match the addition sentence $2 + 2 + 2 + 2 = 8$.

113. Draw an array to match the addition sentence $3 + 3 + 3 + 3 + 3 = 15$

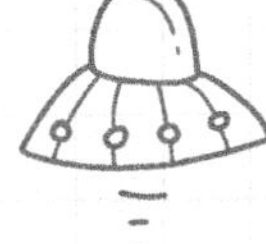

114. Draw an array to match the addition sentence 5 + 5 + 5 = 15.

115. Draw an array to match the addition sentence 4 + 4 + 4 + 4 = 16.

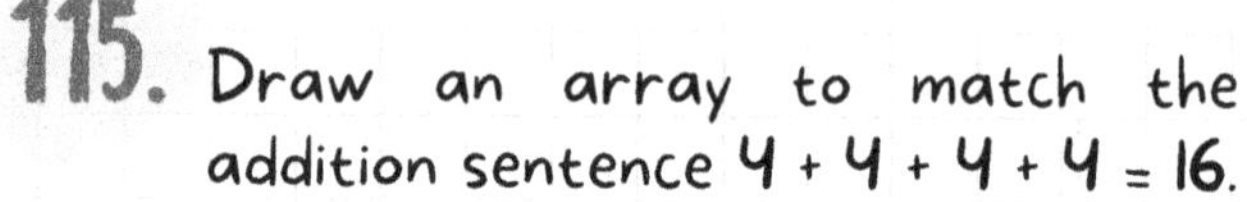

116. Write an addition sentence with equal addends to match the picture. How many legs are on the spiders?

117. Write an addition sentence with equal addends to match the picture. How many fingers are there?

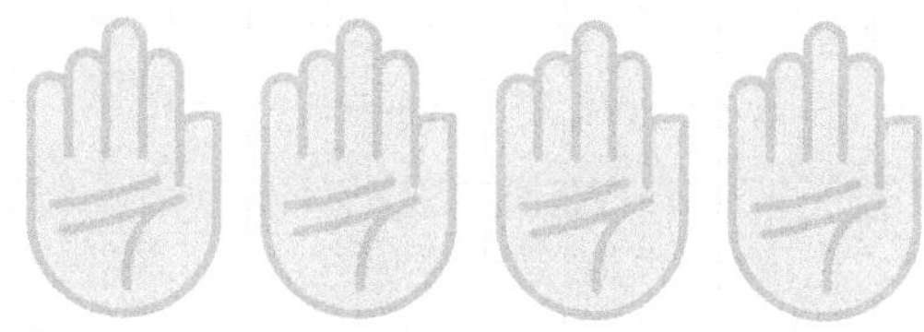

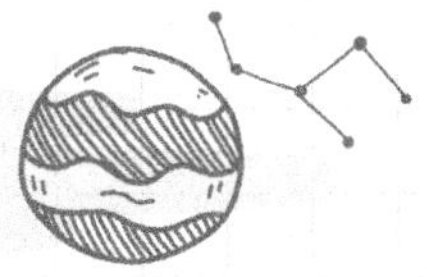

118. Write an addition sentence with equal addends to match the picture. How many feet are there?

119. Write an addition sentence with equal addends to match the picture. How many ears are there?

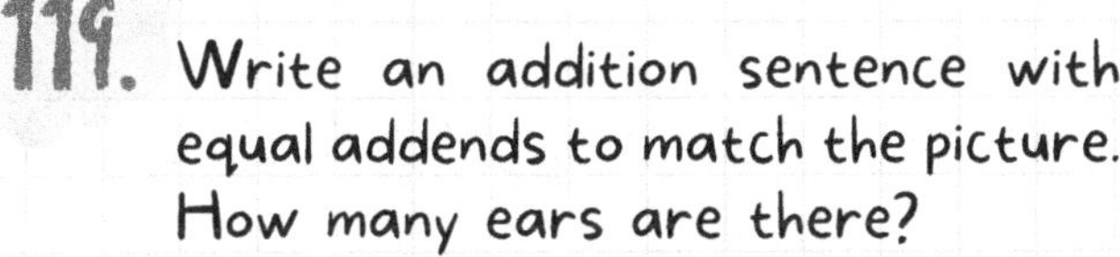

120. Write an addition sentence with equal addends to match the picture. How many eggs are there?

CHAPTER 2

GRADE 2

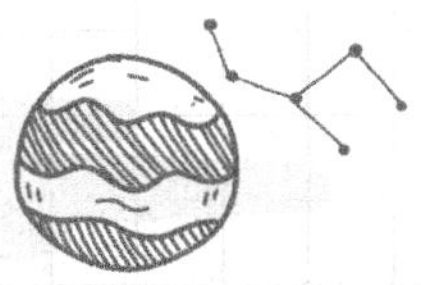

1. How many hundreds are in the number 706?

2. How many tens are in the number 859?

3. How many tens are in the number 632?

4. How many ones are in the number 500?

5. How many hundreds are in the number 498?

6. How many ones are in the number 257?

7. Which group shows the number 40?

A.

C.

B.

D.

TIP: The number of tens rods corresponds to the total number. For example, if there are 7 ten rods then the total is 70.

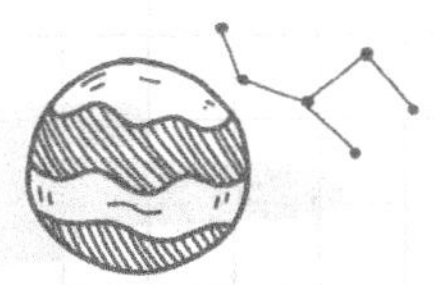

8. Which group shows the number 80?

A.

C.

B.

D.

9. Which group show the number 100?

A. 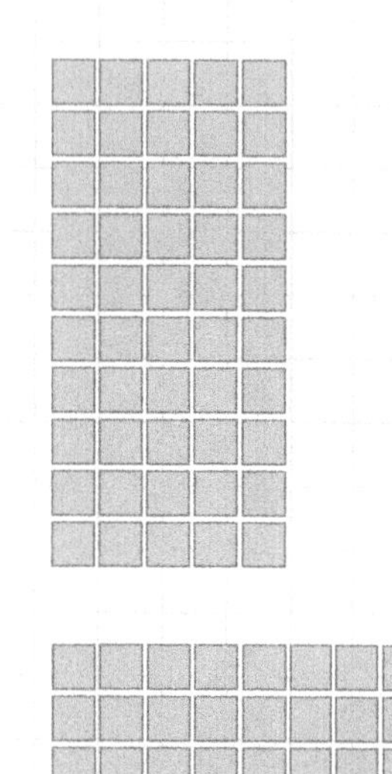

C.

B.

D. 

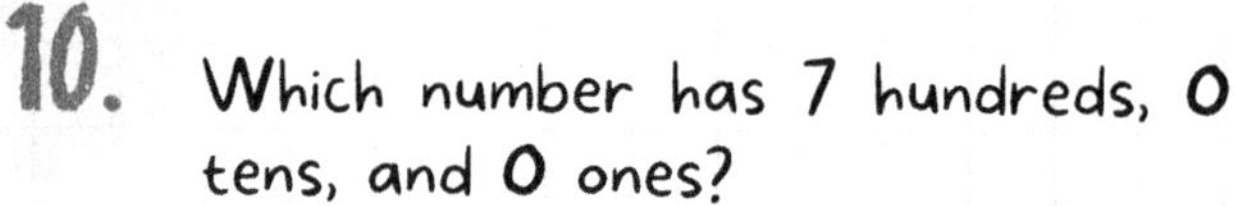

10. Which number has 7 hundreds, 0 tens, and 0 ones?

A. 820
B. 703
C. 400
D. 700

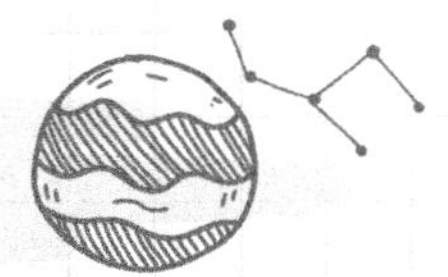

11. Which number has 8 hundreds, 4 tens, and 3 ones?

A. 329
B. 843
C. 384
D. 438

SHOW YOUR WORK

12. Which number has 6 hundreds, 0 tens, and 1 one?

A. 61
B. 601
C. 610
D. 466

SHOW YOUR WORK

13. Write the number that has 3 hundreds, 0 tens, and 0 ones.

SHOW YOUR WORK

14. Write the number that has 4 hundreds, 0 tens, and 0 ones.

SHOW YOUR WORK

15. Write the number that has 9 hundreds, 0 tens, and 0 ones.

16. Write the number that has 7 hundreds, 0 tens, and 0 ones?

17. Start at 35 and count on by 5s.

35, ___ , ___ , ___ , ___ , ___ , ___ , ___

18. Start at 140 and count on by 10s.

140, ___ , ___ , ___ , ___ , ___ , ___ , ___

TIP: When skip counting by 5s, the number in the ones place should either be a 0 or a 5.

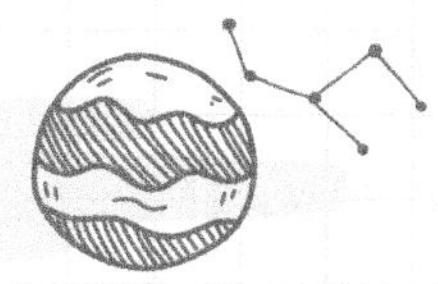

19. Start at 560 and count on by 5s.

560, ___ , ___ , ___ , ___ , ___ , ___ , ___

SHOW YOUR WORK

20. Start at 200 and count on by 100s.

200, ___ , ___ , ___ , ___ , ___ , ___ , ___

SHOW YOUR WORK

21. Start at 850 and count on by 5s.

850, ___ , ___ , ___ , ___ , ___ , ___ , ___

SHOW YOUR WORK

22. Start at 710 and count on by 10s.

710, ___ , ___ , ___ , ___ , ___ , ___ , ___

SHOW YOUR WORK

23. Start at 140 and count on by 100s.

140, ___ , ___ , ___ , ___ , ___ , ___ , ___

24. Which equation shows 679 in expanded form?

A. 679 = 600 + 70 + 9
B. 679 = 700 + 60 + 9
C. 697 = 600 + 90 + 7
D. 766 = 700 + 60 + 6

25. Which equation shows 905 in expanded form?

A. 905 = 900 + 50 + 0
B. 590 = 500 + 90 + 0
C. 905 = 900 + 0 + 5
D. 900 = 900 + 0 + 0

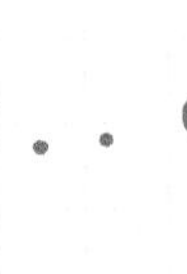

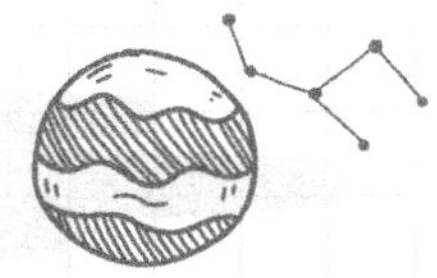

26. Which equation shows **260** in expanded form?

A. 620 = 600 + 20 + 0
B. 265 = 200 + 60 + 5
C. 600 = 600 + 0 + 0
D. 260 = 200 + 60 + 0

27. Write the number **842** in expanded form.

28. Write the number **502** in expanded form.

29. Write the number **198** in expanded form.

30. Compare these two numbers using >, <, and =.

520 ____ 945

31. Compare these two numbers using >, <, and =.

490 ____ 590

32. Compare these two numbers using >, <, and =.

398 ____ 397

33. Compare these two numbers using >, <, and =.

621 ____ 621

34. Compare these two numbers using >, <, and =.

764 ____ 729

TIP: When comparing numbers, first look at the largest place value. If they are equal, go to the next one. Keep going down until you have a number that is greater or less.

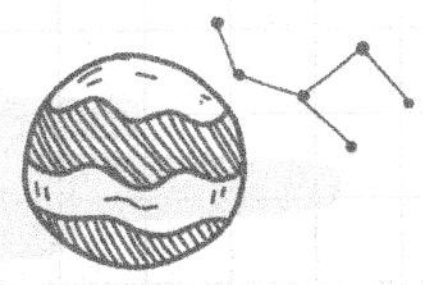

35. Compare these two numbers using >, <, and =.

542 ____ 542

SHOW YOUR WORK

36. Write a number that is greater than 643.

SHOW YOUR WORK

37. Write a number that is equal to 982.

SHOW YOUR WORK

38. Write a number that is less than 201.

SHOW YOUR WORK

39. Write a number that is equal to 399.

SHOW YOUR WORK

40. Write a number that is less than 610.

SHOW YOUR WORK

41. Solve these addition equations.

68 + 32	21 + 65	33 + 60
19 + 40	52 + 38	77 + 21

42. Solve these subtraction equations.

53 - 27	96 - 43	84 - 31
25 - 19	34 - 33	63 - 50

43. Draw a picture using tens rods and units to solve 76 - 29.

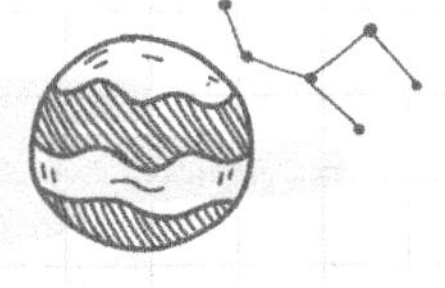

44. Draw a picture using tens rods and units to solve 54 - 36.

45. Draw a picture using tens rods and units to solve 81 - 17.

46. Draw a picture using tens rods and units to solve 66 + 26.

47. Draw a picture using tens rods and units to solve 42 + 55.

48. Solve this addition equation. Show your work.

30 + 15 + 11 + 25 =

SECTION 2: USING PLACE VALUE AND PROPERTIES OF OPERATIONS TO ADD AND SUBTRACT

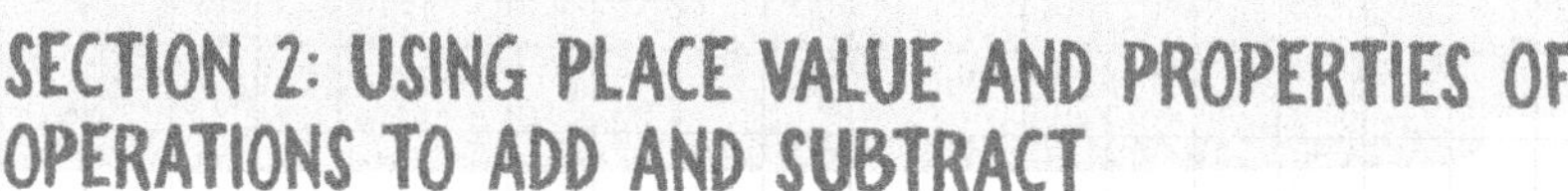

49. Solve this addition equation. Show your work.

51 + 6 + 12 + 20 =

50. Solve this addition equation. Show your work.

26 + 52 + 2 + 19 =

51. Solve this addition equation. Show your work.

38 + 17 + 21 + 5 =

TIP: When adding numbers with multiple addends, add the ones first. Then add the tens. Finally, add the tens and ones together.

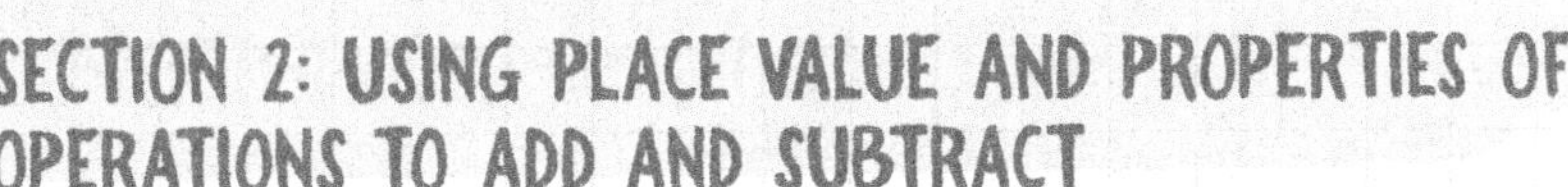

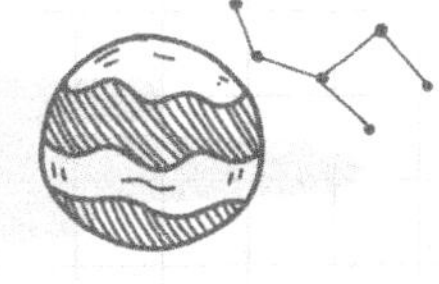

52. What is the correct sum of:

$$\begin{array}{r} 542 \\ +351 \\ \hline \end{array}$$

A. 398
B. 211
C. 893
D. 891

53. What is the correct sum of:

$$\begin{array}{r} 263 \\ +710 \\ \hline \end{array}$$

A. 973
B. 864
C. 379
D. 245

54. What is the correct sum of:

$$\begin{array}{r} 481 \\ +\,255 \\ \hline \end{array}$$

A. 646
B. 637
C. 736
D. 234

55. What is the correct sum of:

$$\begin{array}{r} 569 \\ +\,345 \\ \hline \end{array}$$

A. 224
B. 419
C. 794
D. 914

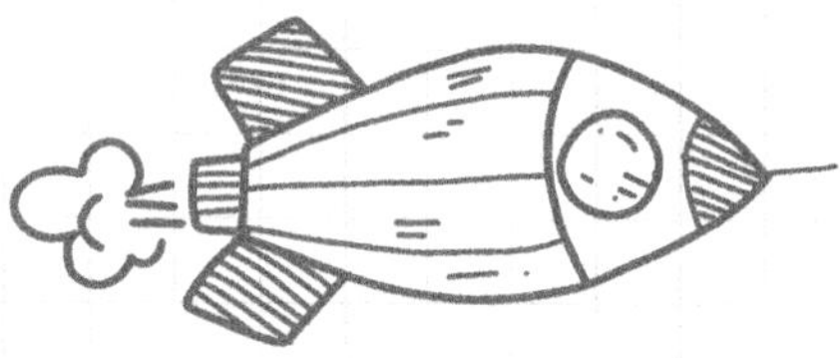

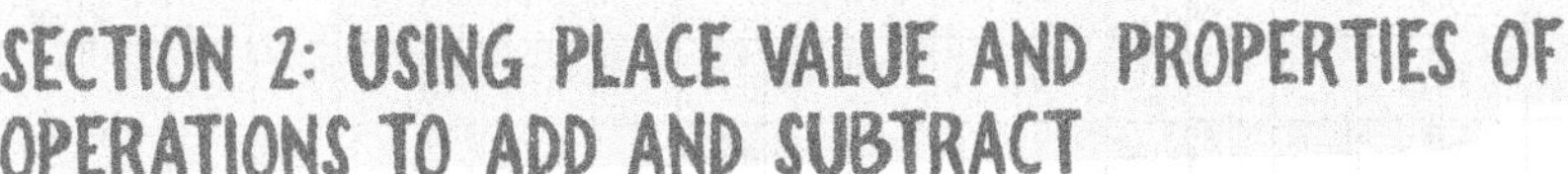

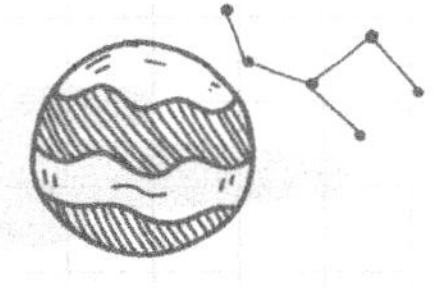

56. Solve the following addition equations:

642 + 261	111 + 601	300 + 456
705 + 194	471 + 222	644 + 110

57. Solve the following addition equations:

597 + 213	327 + 403	685 + 126
476 + 415	618 + 105	700 + 204

58. Find the difference:

$$\begin{array}{r} 367 \\ -152 \\ \hline \end{array}$$

A. 519
B. 512
C. 314
D. 215

59. Find the difference:

$$\begin{array}{r} 935 \\ -710 \\ \hline \end{array}$$

A. 225
B. 224
C. 522
D. 202

TIP: When subtracting, you need to make sure you have enough to take away! If you don't, you need to borrow a ten from the next highest place value.

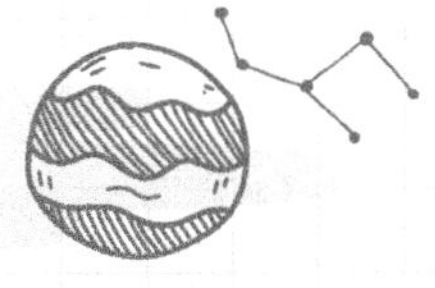

60. Find the difference:

$$\begin{array}{r} 397 \\ -259 \\ \hline \end{array}$$

A. 511
B. 215
C. 138
D. 105

61. Find the difference:

$$\begin{array}{r} 985 \\ -566 \\ \hline \end{array}$$

A. 973
B. 425
C. 421
D. 419

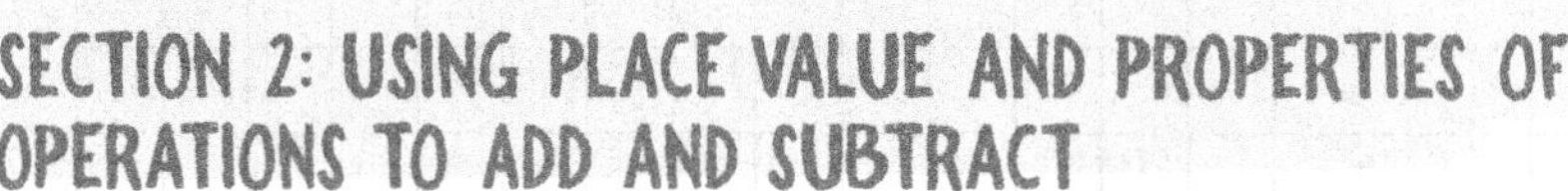

62. Solve the following subtraction equations:

721 - 320	544 - 131	674 - 265
367 - 205	493 - 380	823 - 603

63. Solve the following subtraction equations:

580 - 190	642 - 225	210 - 155
735 - 457	427 - 159	351 - 207

64. Write an equation that represents 10 more than 47. Solve.

65. Write an equation that represents 100 more than 645. Solve.

SHOW YOUR WORK

66. Write an equation that represents 10 less than 278. Solve.

SHOW YOUR WORK

67. Write an equation that represents 100 less than 851. Solve.

SHOW YOUR WORK

68. Write an equation that represents 10 more than 390. Solve.

SHOW YOUR WORK

69. Write an equation that represents 100 less than 529. Solve.

SHOW YOUR WORK

70. Susie's bake shop got an order for **34** cookies for the school, **25** cookies for a birthday party, and **40** cookies for the bake sale. Then the bake sale called and added **100** more cookies. How many cookies did Susie need to make?

71. Emma wanted to buy a new bike. It cost **$200**. She saved up **$123** from doing her chores. How much more money does Emma need to buy the bike?

TIP: Some word problems are more than one step and require you to do more than one operation. Read the problem carefully and look for clues in the text to determine the operation.

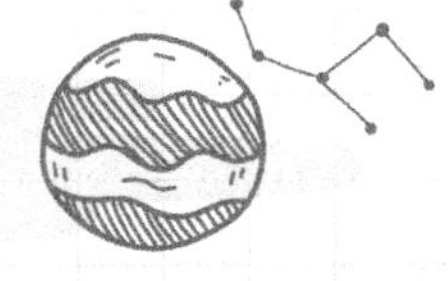

72. For the 100th day of school, Parker wanted to collect 100 pennies. When he counted all he had collected, he found that he only needed 10 more to get to 100 pennies. How many pennies did Parker have?

73. There were 150 apples in the barrel. I bought 24 apples. How many apples were left in the barrel?

74. The art teacher needed to place a supply order. She needed 367 red pom poms and 289 blue pom poms. How many pom poms did the art teacher need?

75. My dad is 40 years old. My grandpa is 72 years old. How much older is my grandpa than my dad?

76. I took my dog to the vet, and he weighs 58 pounds. The next time we went back, the vet said we need to put him on a diet because he gained 10 pounds. How much did my dog weigh?

77. Jess and Dee were fighting over who had more dolls. Jess thought she had more because she had 84, and Dee thought she had more because she had 97. Who had more, and how many more did she have?

78. Natalie was collecting stamps. She got 19 from her grandma, 37 from her aunt, 28 from her cousin, and 5 from her mom. How many stamps did Natalie have in all?

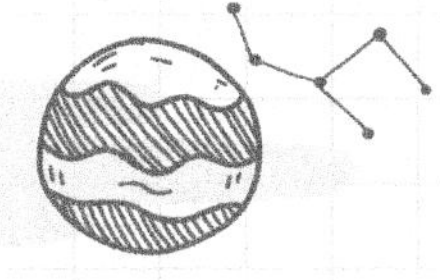

79. My little brother spilled a box of **64** crayons on the ground. After we picked them all up, we only had **57**. How many crayons were still missing?

80. There were **735** seats in the soccer stadium. On the day of the big game, they sold **729** tickets. David and his family wanted to buy **5** tickets to see the game. Are there enough tickets for his family to go to the game?

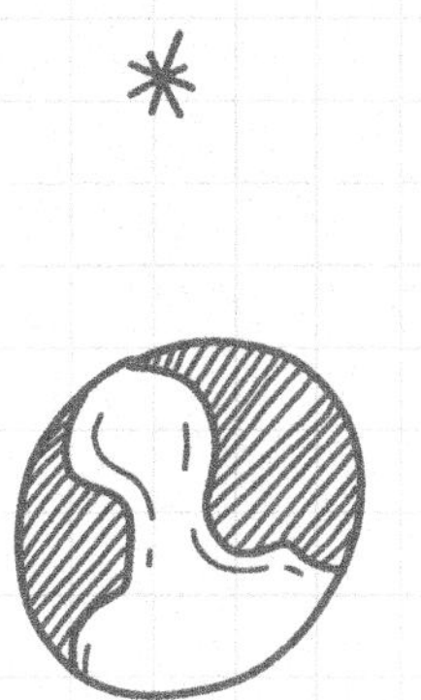

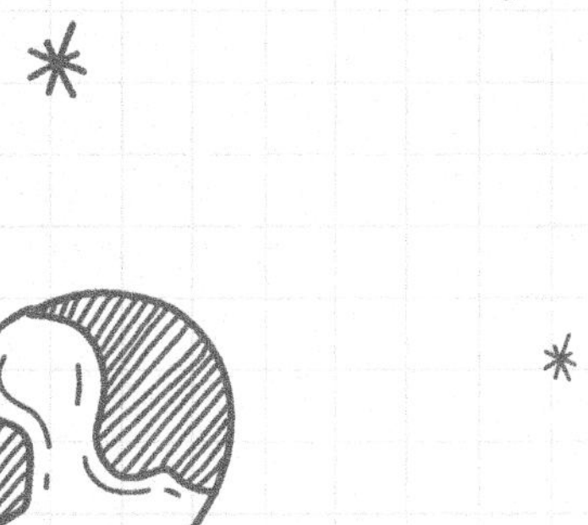

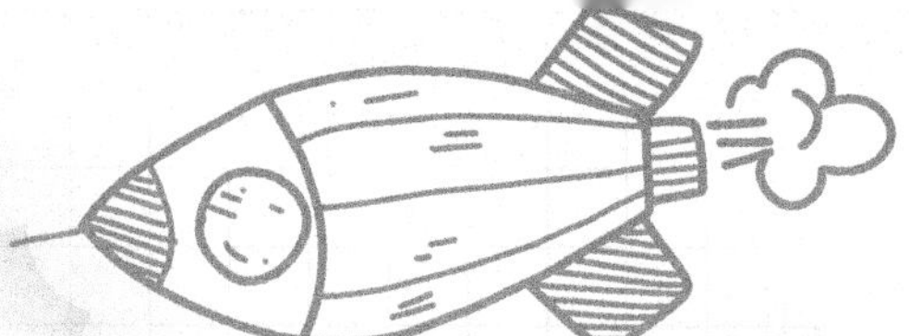

CHAPTER 3

GRADE 2

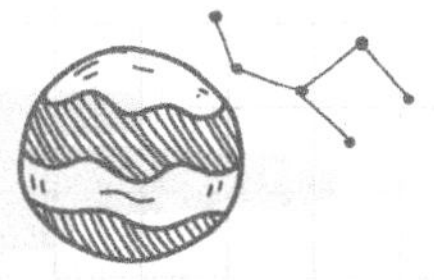

1. What would be the most efficient tool to measure the length of a couch?

A. A ruler
B. A tape measure
C. A measuring cup
D. A yardstick

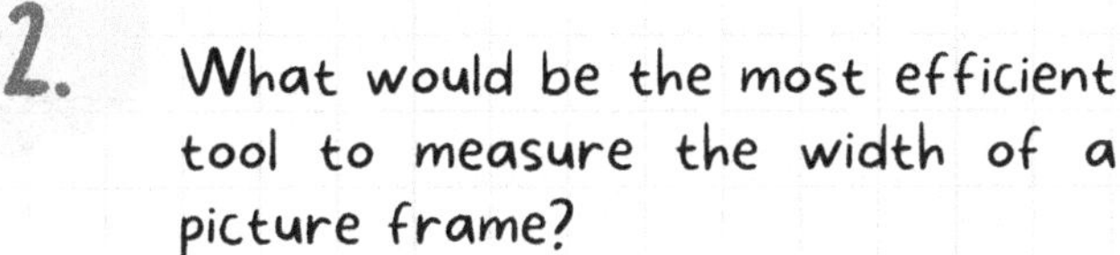

2. What would be the most efficient tool to measure the width of a picture frame?

A. A yardstick
B. A meter stick
C. A ruler
D. A measuring cup

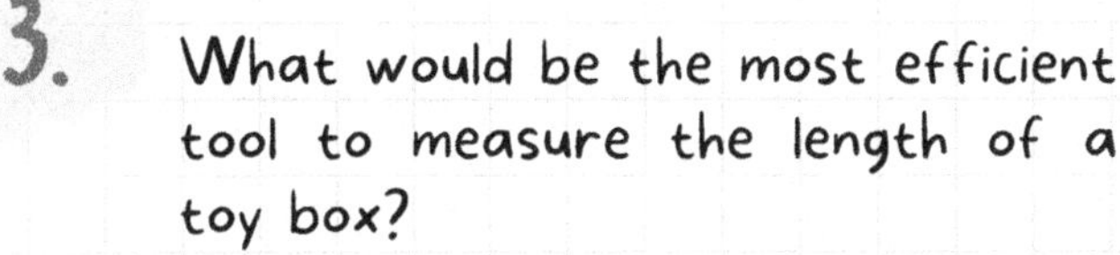

3. What would be the most efficient tool to measure the length of a toy box?

A. A measuring cup
B. A ruler
C. A yardstick
D. Thermometer

4. To measure the length of a book, what standard unit would be the most efficient?

A. feet
B. inches
C. yards
D. pounds

5. To measure the height of a door, what standard unit would be the most efficient?

A. centimeters
B. inches
C. pounds
D. feet

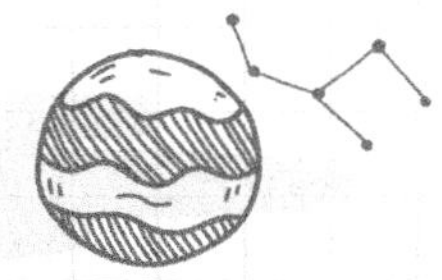

6. Measure the shark to the closest inch.

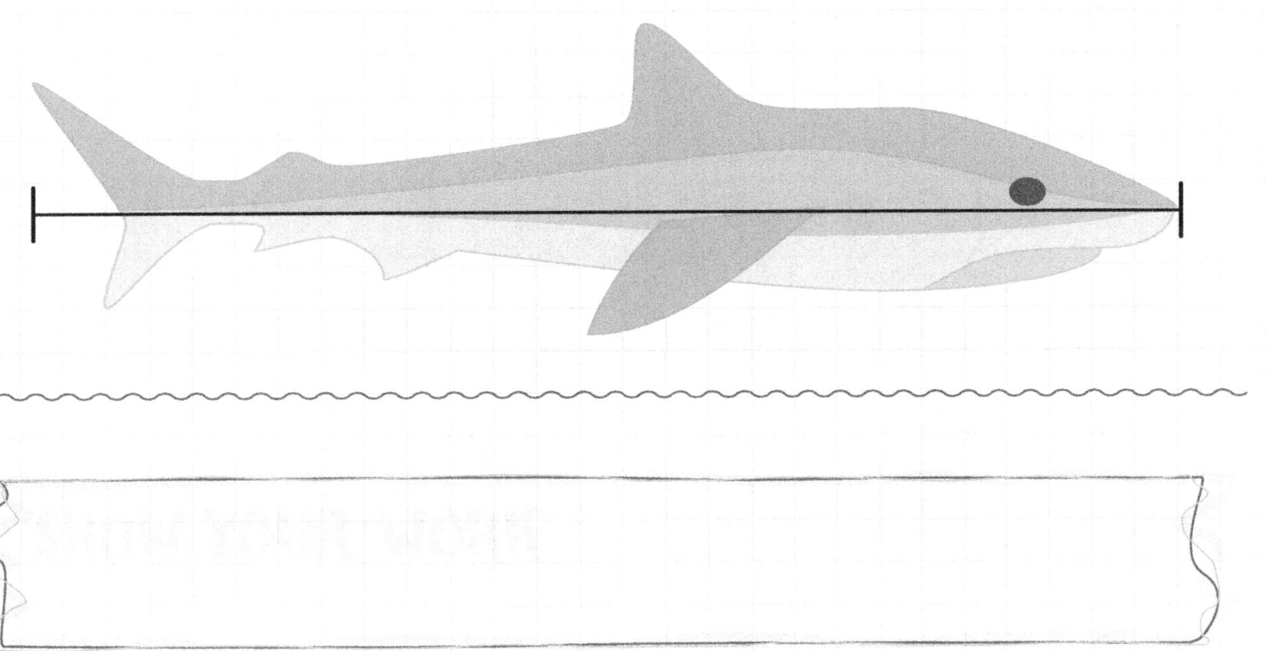

7. Measure the caterpillar to the closest centimeter.

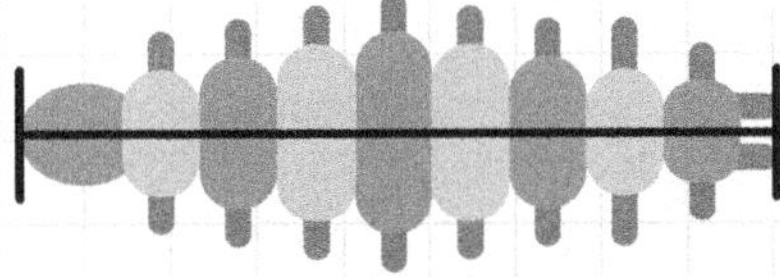

8. Measure the fish to the closest centimeter.

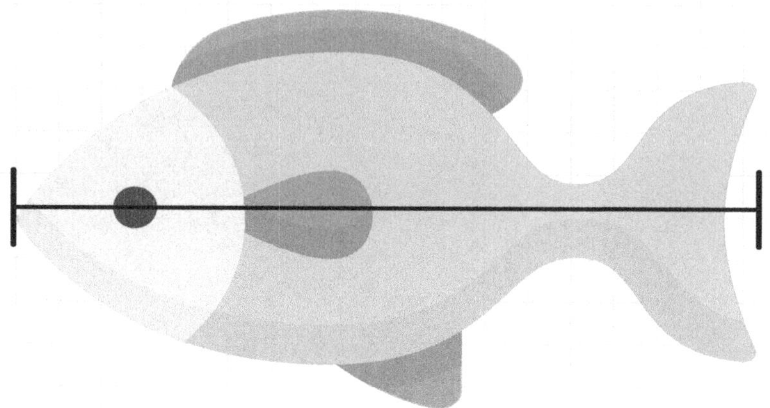

TIP: Make sure the end of the ruler is lined up to the edge of the object you are measuring. If it is not lined up, your measurements will be inaccurate.

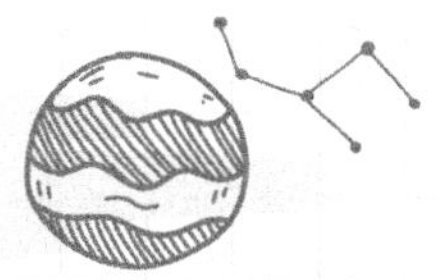

9. Measure the pillow to the closest inch.

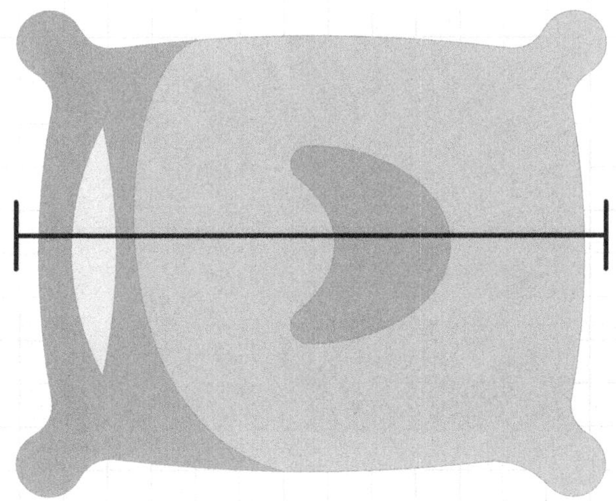

10. Measure the sneaker to the closest centimeter.

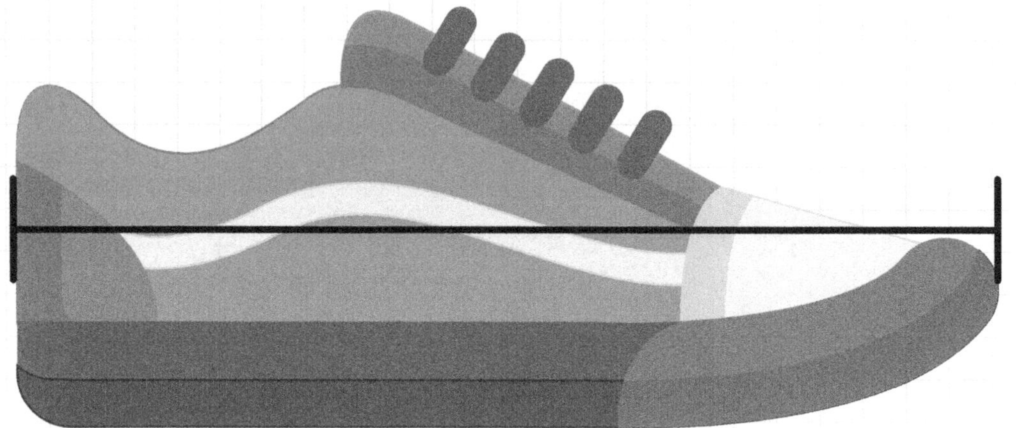

11. The table measures 48 inches, or 36 paperclips. How many more inches than paper clips are needed to measure the table?

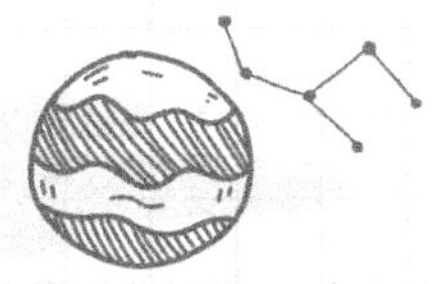

12. The television measures 65 inches, or 50 paper clips. How many fewer paper clips than inches are needed to measure the table?

13. The baseball bat measures 106 centimeters, or 18 toothpicks. How many more centimeters than toothpicks are needed to measure the baseball bat?

14. The rug measures 152 centimeters, or 26 toothpicks. How many fewer toothpicks than centimeters are needed to measure the rug?

15. I am 4 feet tall or 6 tiles. How many fewer feet than tiles are needed to measure me?

16. Lindsey is 5 feet tall, or 8 tiles. How many more tiles than feet are needed to measure Lindsey?

17. Joe measured his book in inches, and it measured 12 inches. Drew measured the same book with paper clips and got 9 paperclips. Why did they get different results when measuring the same book?

A. The paper clips are bigger than an inch.
B. The paper clips and inches are equal.
C. The paper clips are smaller than an inch.
D. The paper clips were spaced too far apart.

TIP: It takes more smaller units to measure an object and few longer units to measure the same object.

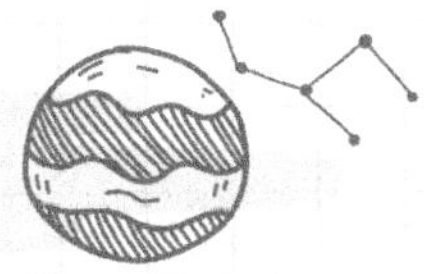

18. Pam measured her height in inches and measured 48 inches. Then she measured herself in centimeters and measured 121 centimeters. Why did Pam get different measurements when measuring her height?

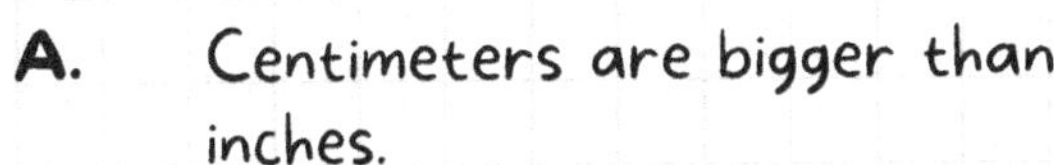

A. Centimeters are bigger than inches.
B. Inches are smaller than centimeters.
C. Centimeters are smaller than inches.
D. Inches and centimeters are equal.

19. Jim measured his bed with a measuring tape, and it was 6 feet long. When he measured it with a ruler, it was 72 inches. Why did Jim get different measurements when measuring his bed?

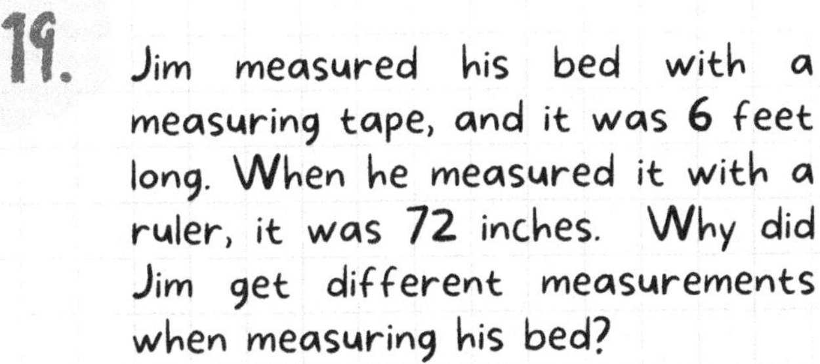

A. Inches are bigger than feet.
B. Feet are bigger than inches.
C. Feet are smaller than inches.
D. Feet and inches are equal.

20.

Kelly's bedroom is 3 yards wide. When she measured it when a ruler, she measured 18 feet. Why did Kelly get different measurements when measuring the width of her room?

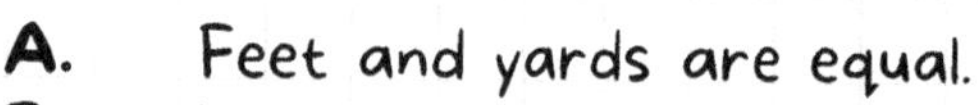

A. Feet and yards are equal.
B. Yards are smaller than feet.
C. Feet are bigger than yards.
D. Yards are bigger than feet.

21.

Estimate the length of the shorter fence.

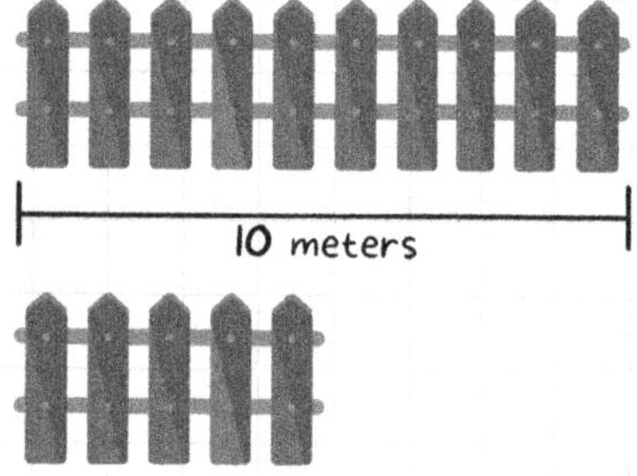

A. 15 meters
B. 5 meters
C. 3 feet
D. 24 inches

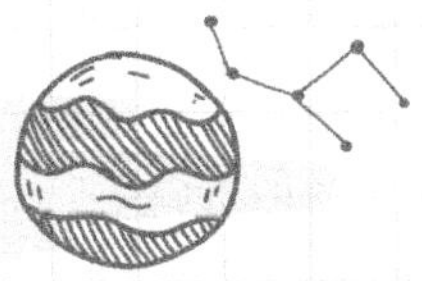

22. Estimate the length of the longer bed.

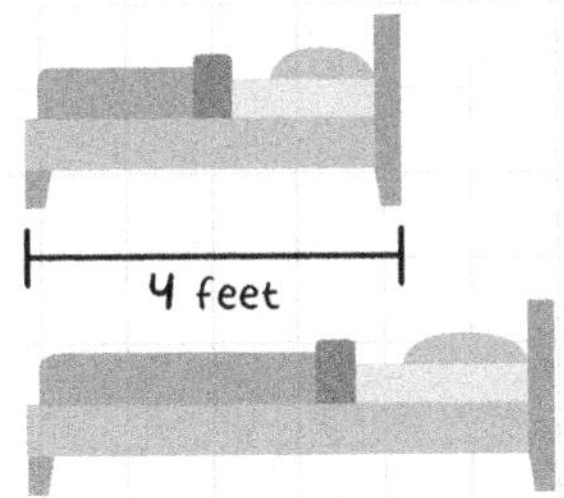

A. 2 feet
B. 10 feet
C. 6 feet
D. 1 meter

23. Estimate the height of the taller door.

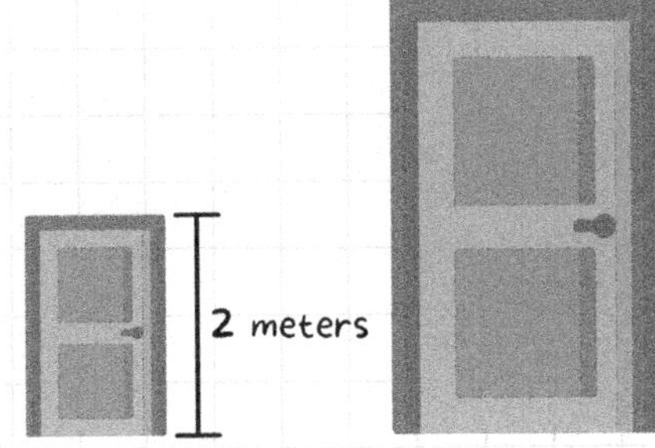

A. 1 meter
B. 4 meters
C. 4 feet
D. 4 yards

24. Estimate the width of the longer bench.

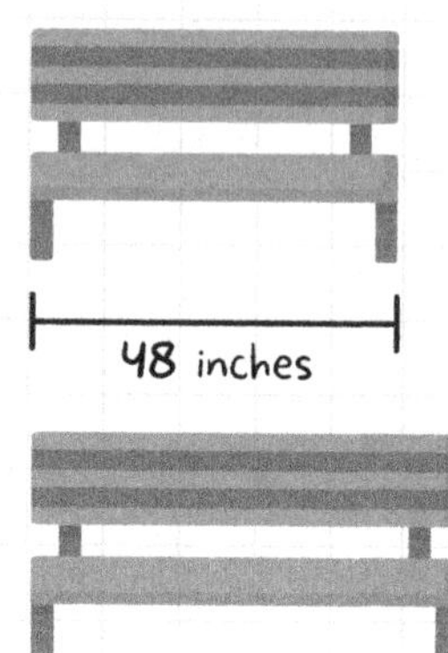

A. 30 inches
B. 4 feet
C. 1 meter
D. 60 inches

TIP: When estimating a measurement, try to decide if the unknown length is much bigger or smaller. Using a known measurement will help you be more accurate.

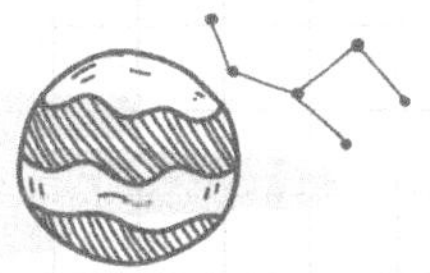

25. Estimate the height of the shorter eraser.

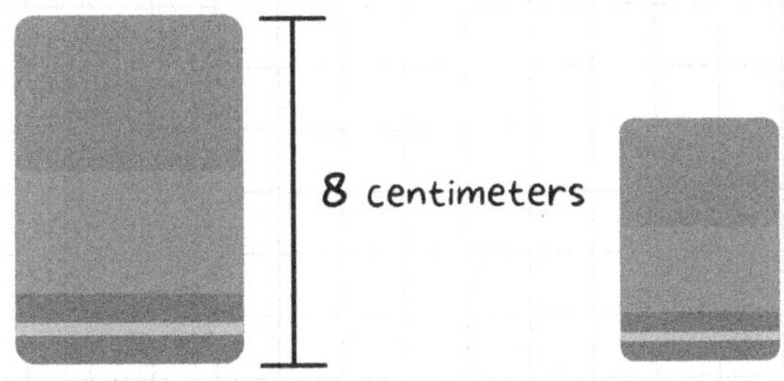

A. 1 centimeter
B. 10 centimeters
C. 5 centimeters
D. 2 inches

26. Estimate how long a cell phone may be.

A. About 6 inches
B. About 3 centimeters
C. About 1 foot
D. About 1 yard

27. Estimate how long a crayon may be.

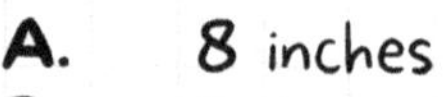

A. 8 inches
B. 2 feet
C. 4 centimeters
D. 1 meter

28. Estimate how tall a basketball hoop may be.

A. 10 feet
B. 12 inches
C. 5 yards
D. 20 centimeters

29. Estimate how wide a dollar bill may be.

A. 3 feet
B. 5 meters
C. 7 centimeters
D. 10 inches

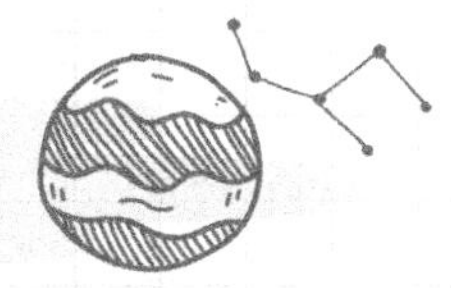

30. Estimate how many meters long a pool may be.

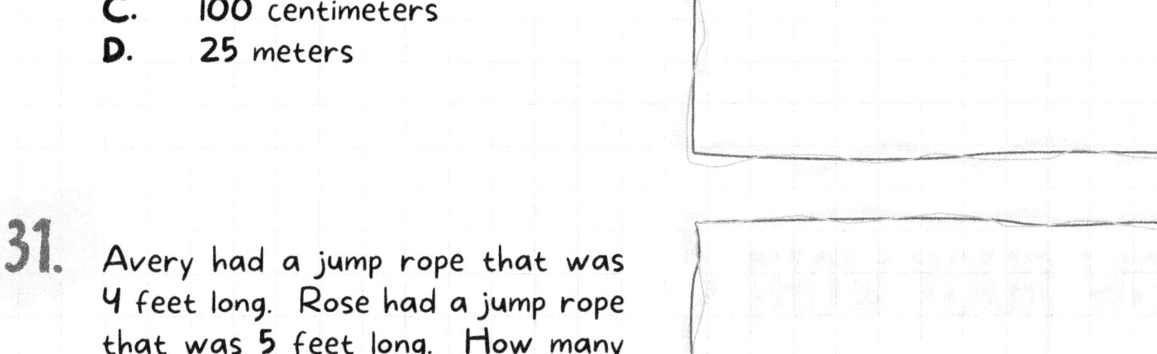

A. 4 feet
B. 50 inches
C. 100 centimeters
D. 25 meters

31. Avery had a jump rope that was 4 feet long. Rose had a jump rope that was 5 feet long. How many feet longer is Rose's jump rope than Avery's?

32. Justin and Jimmy are on the track team. Justin is running in the 50 yard dash. Jimmy is running in the 25 yard dash. How many more yards is Justin running than Jimmy?

33. My dad is building a bench for my mom's garden. My brother and I asked to help him. My dad asked me to cut a board 12 feet long. My brother cut a board that was 9 feet long. How much longer was my board than my brother's?

34. Jackie needs to buy carpet for her bedroom. Her bedroom is 6 yards. She bought 13 yards of carpet. How much extra carpet will Jackie have left over?

TIP: When comparing the lengths of objects, use a subtraction equation to determine how much longer or shorter an object may be compared to another. Make sure you start with the size of the bigger object.

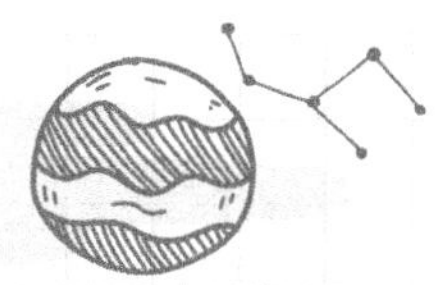

35. Nate has to carry his couch into his new house. The width of the couch is 60 inches. His front door is 48 inches wide, and his basement door is 72 inches wide. Which door should he use, and how much extra room will be have to get through the door?

36. I got my dog, Sadie, a new leash. Her old leash was too short. It was 119cm. Her new leash is 182cm. How much longer is the new leash than the old leash.

37. My teacher asked me to cut her 24 inches of yarn. I made a mistake, and I cut too much. I have 30 inches of yarn. How much yarn do I need to trim off so I can give my teacher the 24 inches that she asked for?

38. Jill's new pencil was 18 centimeters long when she got it. Over the week, she sharpened it 5 times. By Friday, it was 4 centimeters long. How much of her pencil did she sharpen away?

39. Our kitchen table is 64 inches long. The tablecloth is 100 inches long. How much longer is the tablecloth than the table?

40. My mom and I planted sunflowers in our garden. My sunflower grew 84 inches, and my mom's sunflower grew 76 inches. How much bigger is my sunflower than my mom's sunflower?

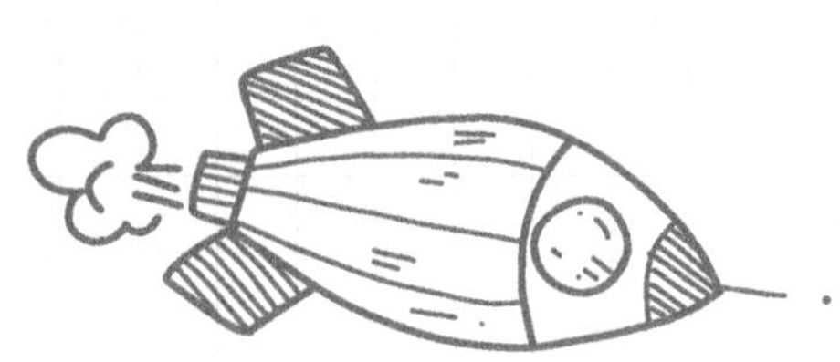

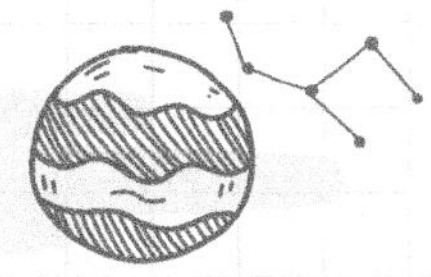

41. Lynn's flower garden is 38 feet from her house. Her hose can reach past her flower garden by 10 feet. How long is Lynn's hose?

42. I walked 6 yards to my mailbox to get the mail. My neighbor walked 10 yards down his driveway to get the mail. How much farther did my neighbor walk than me?

43. In gym class, we were going to play tug of war. We needed a rope that was 15 feet. Mr. Smith, the gym teacher, had a 10 foot rope. Mrs. MacDonald, the principal, had another rope. When we tied them together, it was 15 feet, and we could play tug of war. How long was Mrs. MacDonald's rope?

44. For Ellen's birthday party, she needed 4 tables that were 10 feet long each. If she put all the tables side by side, what would be the length? Draw a picture and write an equation to solve.

45. A full grown giraffe can be about 18 feet tall. When a baby giraffe is born, it is only 6 feet tall. How much does a giraffe grow from baby to adult?

46. Hannah needed to paint her fence. She painted 14 feet of her fence blue, but then she ran out of paint. She still had another 25 feet to go. What is the total length of Hannah's fence?

TIP: Drawing a picture can help you solve a problem. Make sure you draw an accurate picture with details and labels!

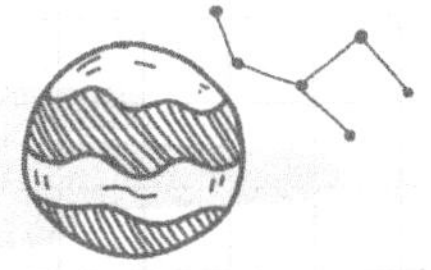

47. Lori wanted to knit a blanket. She needed 210 yards of yarn. She wanted some of the blanket to be yellow and some of the blanket to be pink. Write an equation to show how much could be yellow and how much could be pink.

48. A red kangaroo can jump 25 feet in a single leap. A hare can jump 10 feet in one jump. How much farther can a kangaroo jump compared to a hare?

49. My class lined up for lunch. Our line was 32 feet long. The girls measured 18 feet in line. How long did the boys measure in line?

50. At the beach, we got a 12 inch hot dog to share. I ate some of it, and then gave it to my sister. When she got the hot dog, it was 5 inches. How many inches did I eat?

51. Frank needed to put a fence around his vegetable garden so the animals don't eat his vegetables. The fence was 28 feet all around. How long was the right side of the fence?

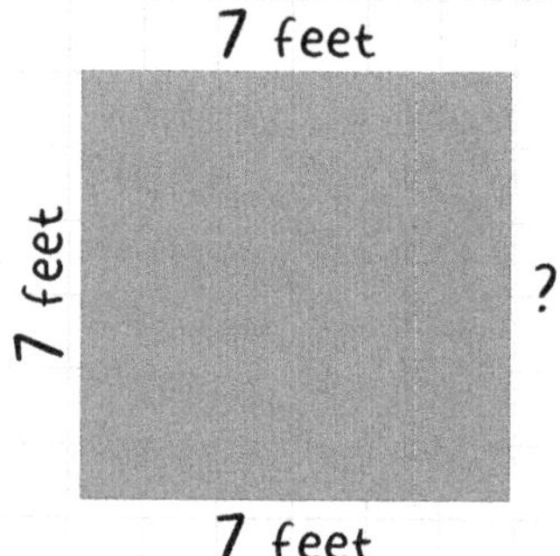

52. Draw a line that is 6 inches long. 4 of the inches should be red, and the rest should be blue. Write an equation to show the total.

53. Draw a line that is 5 inches long. 1 of the inches should be red, and the rest should be blue. Write an equation to show the total.

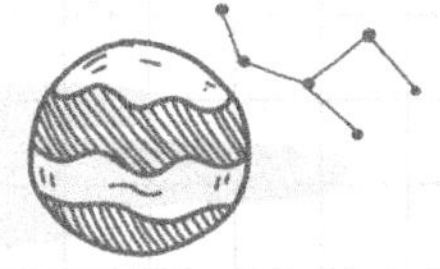

54. Draw a line that is 5 inches long. 3 of the inches should be red, and the rest should be blue. Write an equation to show the total.

55. Here is a line that is 6 inches long. Draw a line across it and cut it at 2 inches. How much is left? Write an equation to solve.

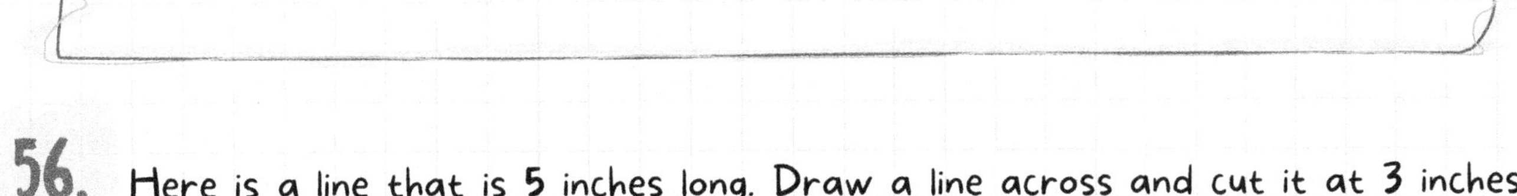

56. Here is a line that is 5 inches long. Draw a line across and cut it at 3 inches. How much is left? Write an equation to solve.

57. Draw a picture to solve. 1 car is 13 feet. How many cars would you have if the total was 39 feet?

58. Draw a picture to solve. 1 box is 14 inches across. How long would it be if you lined up 4 boxes?

59. Draw a picture to solve. 1 spoon is 18 centimeters. How many spoons would you have if the total was 36 centimeters?

60. Draw a picture to solve. 1 bat is 3 feet long. How long would it be if you lined up 10 bats?

TIP: When solving equations with repeated addends, use your double facts to help you.

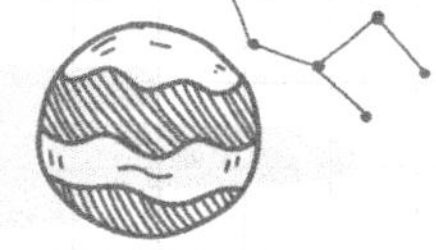

61. Fill in the missing numbers on the numberline.

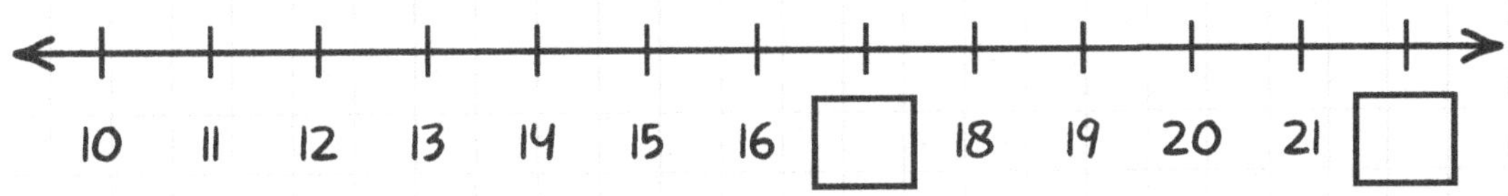

62. Fill in the missing numbers on the numberline.

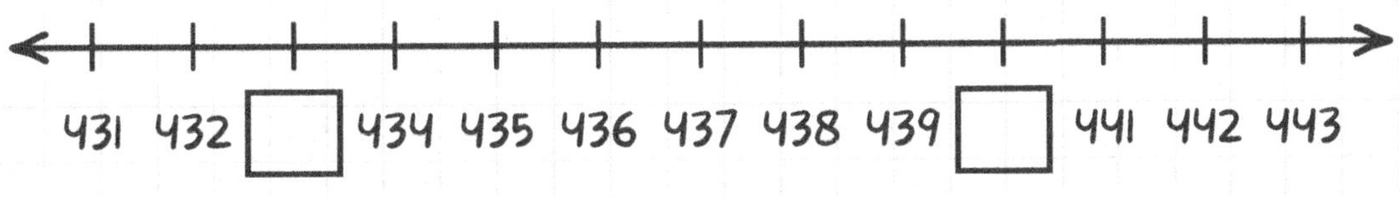

63. Fill in the missing numbers on the numberline.

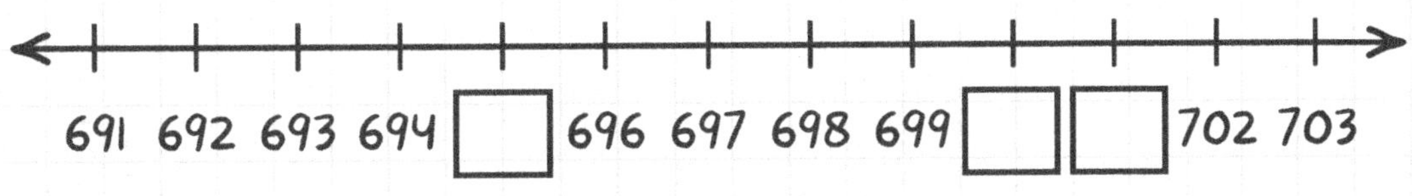

64. Fill in the missing numbers on the numberline.

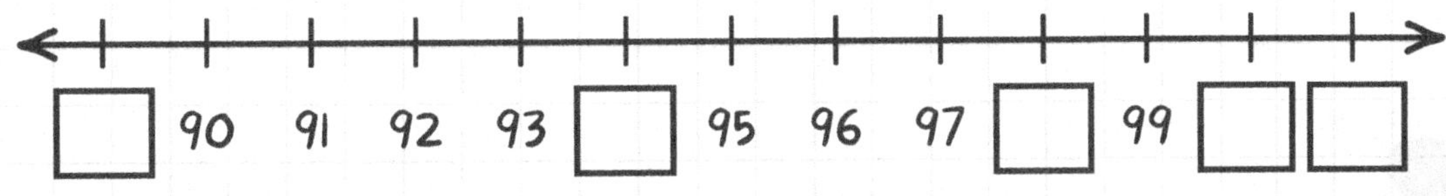

65. Place the number **25** on the number line below.

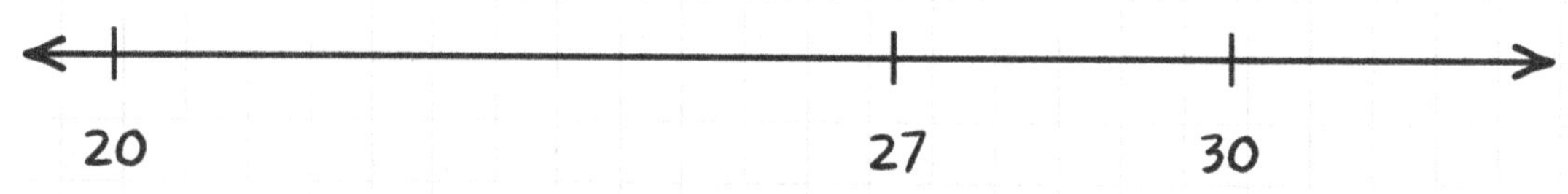

66. Place the number **301** on the number line below.

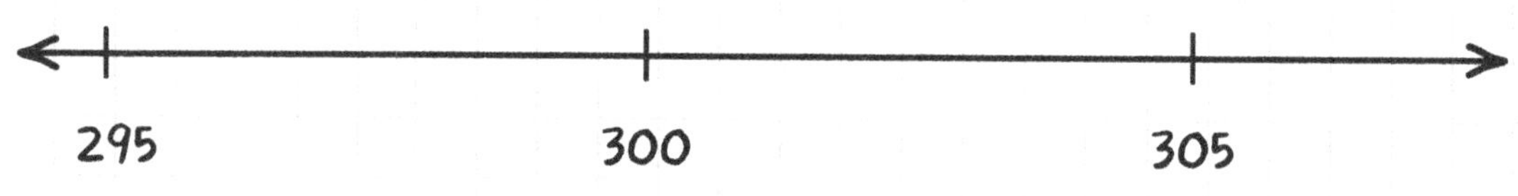

67. Place the number **65** on the number line below.

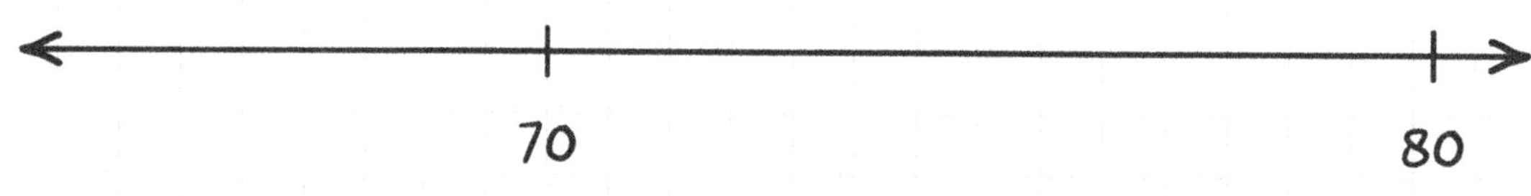

68. Place the number **867** on the number line below.

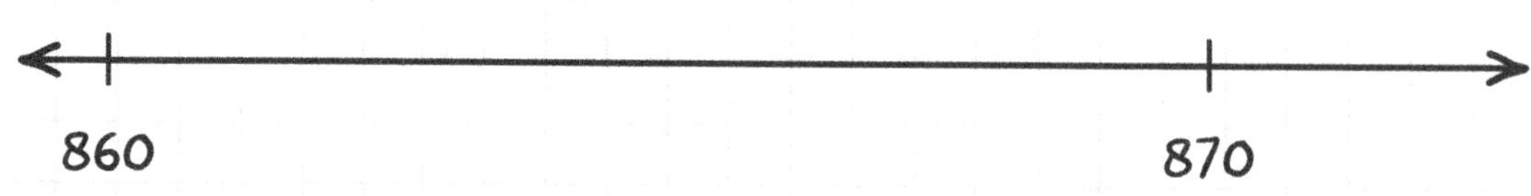

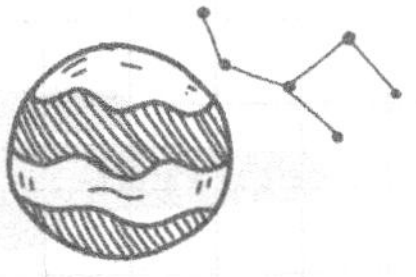

69. Show how to solve this equation on the number line: 19 + 7 = ?

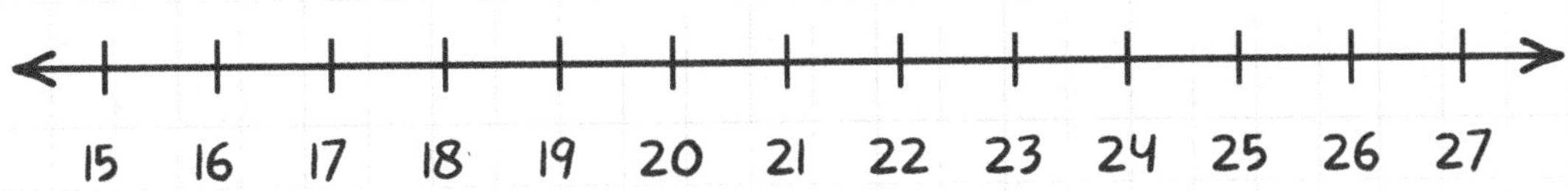

70. Show how to solve this equation on the number line: 274 + ? = 286

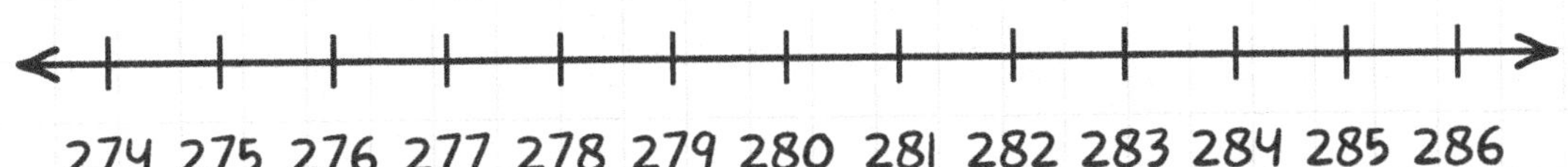

71. Show how to solve this equation on the number line: 621 - 8 = ?

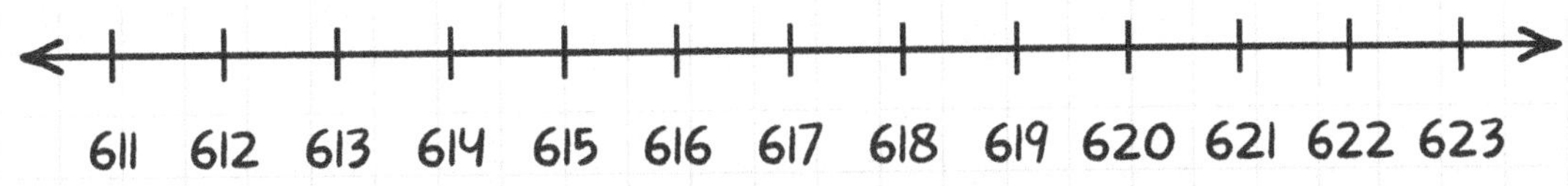

72. Show how to solve this equation on the number line: 501 - ? = 490

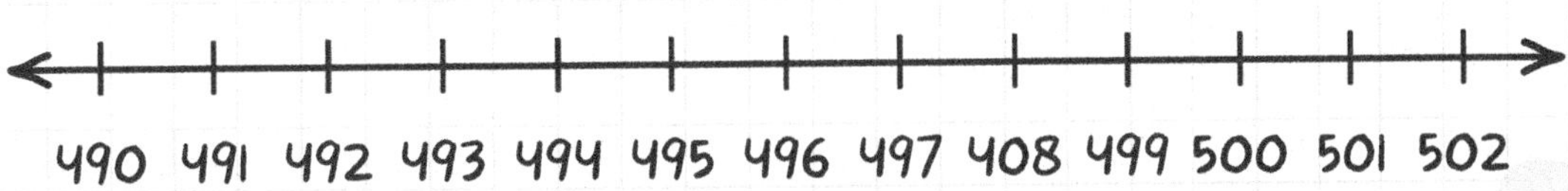

73. Write the addition equation that the number line depicts.

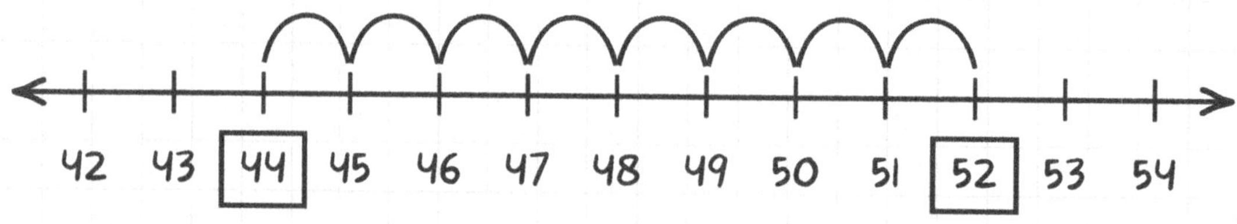

74. Write the addition equation that the number line depicts.

TIP: When placing numbers on a number line, start at the closest number and count on or back to estimate where the number belongs.

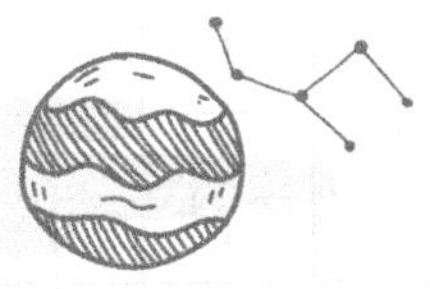

75. Write the subtraction equation that the number line depicts.

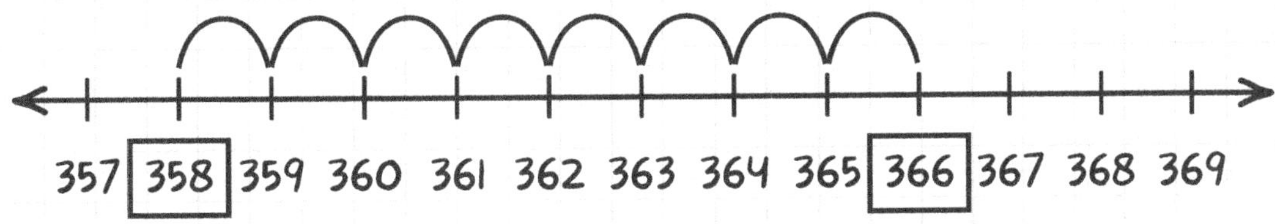

76. Write the subtraction equation that the number line depicts.

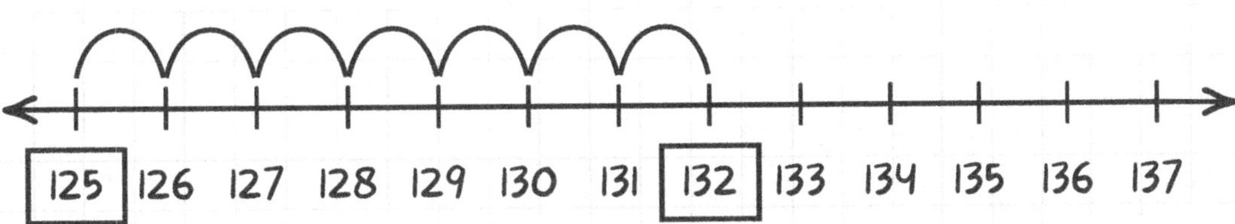

77. Solve the subtraction equation by showing your work on the number line:
62 - 25 = ?

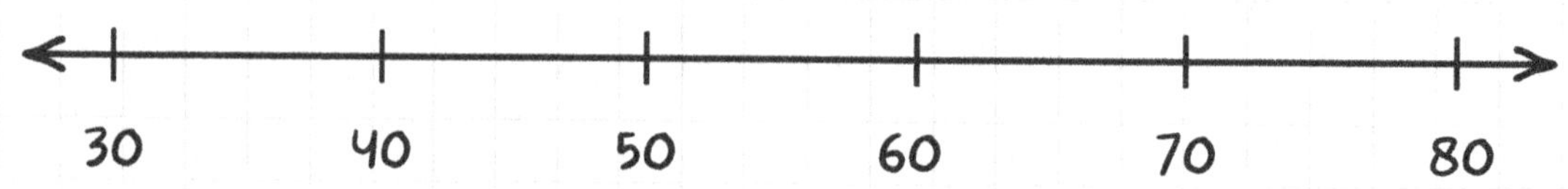

78. Solve the subtraction equation by showing your work on the number line:
325 - 58 = ?

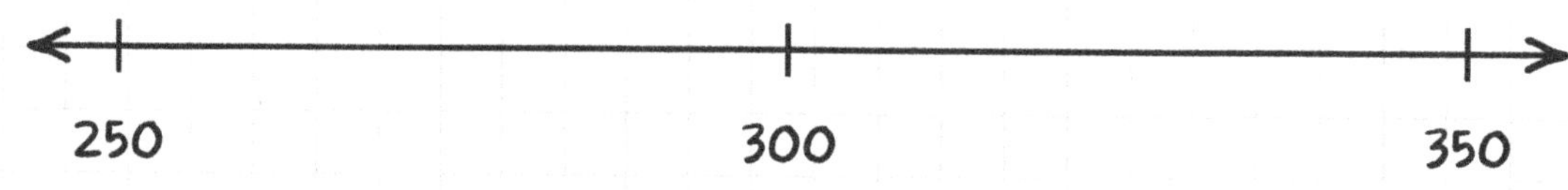

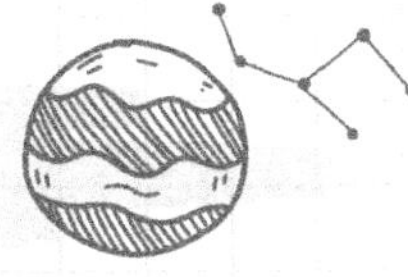

79. Solve the addition equation by showing your work on the number line: 86 + 102 =?

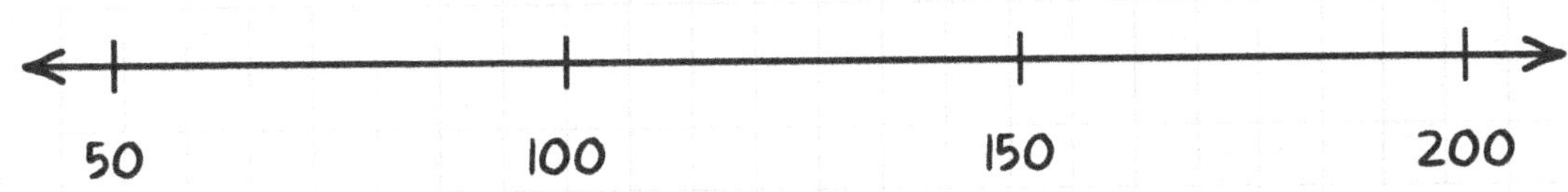

80. Solve the addition equation by showing your work on the number line: 612 + 99 = ?

81. What time does the clock show?

A. 4:00
B. 4:20
C. 4:12
D. 12:20

TIP: When counting on a number line, make sure you are always moving forward or backward. Do not count in place.

82. What time does the clock show?

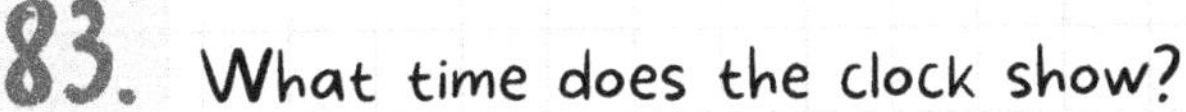

A. 2:35
B. 7:20
C. 7:10
D. 2:07

83. What time does the clock show?

A. 9:45
B. 9:40
C. 8:50
D. 8:45

84. What time does the clock show?

A. 1:30
B. 6:00
C. 6:05
D. 12:30

85. What time does the clock show?

A. 4:50
B. 10:15
C. 10:20
D. 4:10

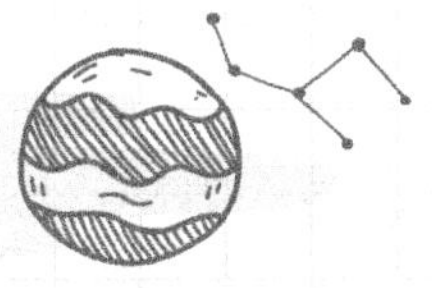

86. Show 2:30 on the clock below:

87. Show 6:15 on the clock below:

88. Show 11:00 on the clock below:

89. Show half past 3 on the clock below:

90. Show quarter of 5 on the clock below:

TIP: When reading a clock, the hour hand is shorter and you read that first. When the minute hand is on the 12, a new hour is starting.

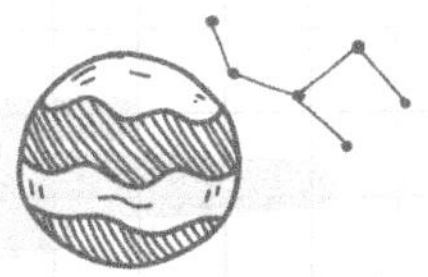

91. What time does this clock show:

04 : 30

A. Quarter past 4:00
B. Half past 4:00
C. Half past 3:00
D. Quarter of 4:00

92. What time does this clock show:

05 : 15

A. Half past 5:00
B. Quarter of 5:00
C. Quarter past 5:00
D. Quarter past 1:00

93. What time does this clock show:

12 : 55

A. Quarter past 12:00
B. Half past of 12:00
C. 5 of 1:00
D. 5 after 12:00

94. What time does this clock show:

02 : 05

A. 5 after 2:00
B. Quarter past 2:00
C. 5 of 2:00
D. Quarter of 2:00

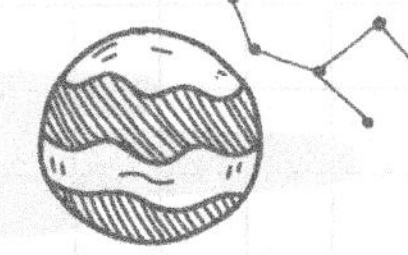

95. What time does this clock show:

01 : 45

A. Quarter past 1:00
B. Half past 1:00
C. Quarter of 2
D. 5 of 1:00

SHOW YOUR WORK

96. Carly woke up for school at 7:00am. She took a half hour to get dressed and a half hour to eat breakfast. Then she got right on the bus. What time did Carly get on the bus?

A. 7:30am
B. 8:00pm
C. 9:00am
D. 8:00am

SHOW YOUR WORK

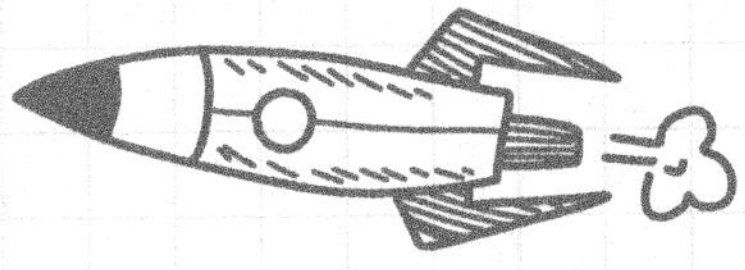

97. The train runs from Philadelphia to New York every 15 minutes. If Nina missed the 12:00pm train, what is the next train she could get on to get to New York from Philadelphia?

A. 12:00am
B. 11:45am
C. 12:15pm
D. 12:30pm.

98. Justin has a doctor's appointment at 3:00pm and a dinner reservation at 5:00pm. How much time does Justin have between the doctor and dinner?

A. 3 hours
B. 1 hour
C. A half hour
D. 2 hours

TIP: If the minute hand is on the 6 or higher, the hour hand should be halfway or closer to the next hour.

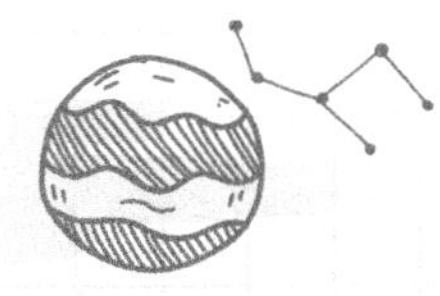

99. Jessie has dance class at 4:15pm. Her class is an hour and a half. What time will her class be over?

A. 5:15pm
B. Quarter of 6:00
C. Half past 5:00
D. 7:00pm

100. David has a flight at 5:00pm. The flight is 6 hours and 10 minutes. What time will David land?

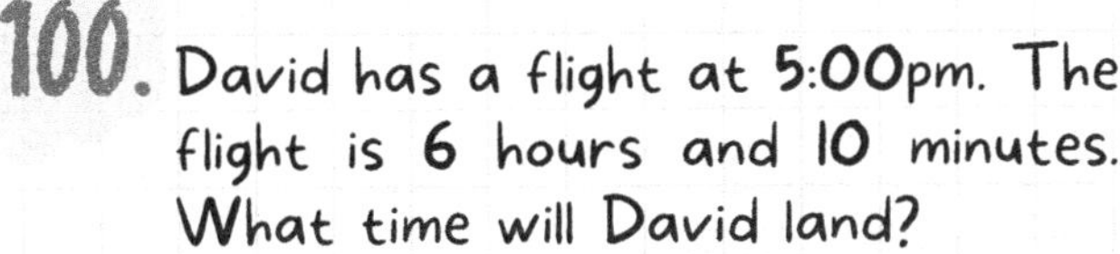

A. 11:10pm
B. 10:15pm
C. 11:10am
D. Quarter after 11

101. Count the money below.

A. 50¢
B. 48¢
C. 45¢
D. 35¢

102. Count the money below.

A. 89¢
B. 30¢
C. 24¢
D. 29¢

103. Count the money below.

A. 85¢
B. 90¢
C. 75¢
D. 83¢

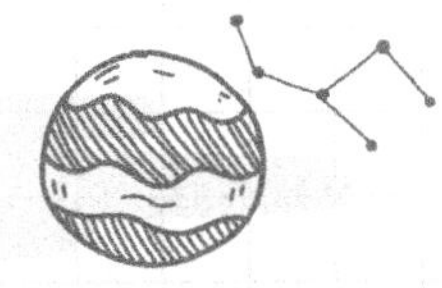

104. Count the money below.

A. 51¢
B. 16¢
C. 31¢
D. 76¢

105. Count the money below.

A. 90¢
B. 55¢
C. $1
D. 60¢

106. Mary has this much money:

She earned **50¢** more by doing her chores. How much money does Mary have now?

A. 85¢
B. 88¢
C. 38¢
D. 63¢

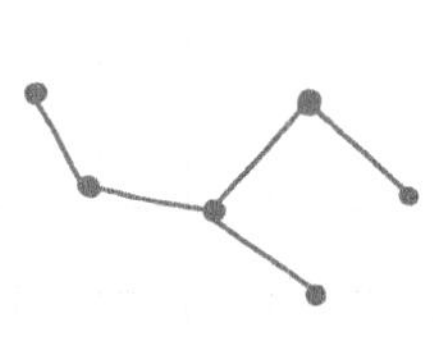

TIP: You get change back if you give too much money at a store! To figure out how much change you get back, subtract the amount you gave with the price of your purchase.

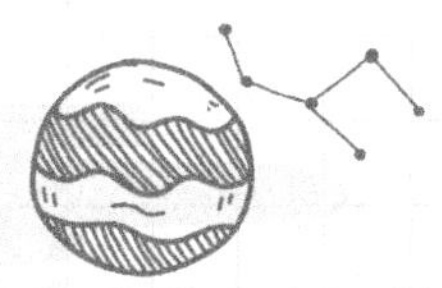

107. Anna saved up this much money:

Her brother asked to borrow **25¢** play a video game at the arcade. How much money does Anna have left?

A. 65¢ **C.** 70¢
B. 100¢ **D.** 95¢

108. Matt has this much money in his piggy bank:

He spends **50¢** on a pretzel for snack. How much money does Matt have left in his piggy bank?

A. 92¢ **C.** 17¢
B. 67¢ **D.** 42¢

109. Charlie has this much money in his pocket:

He found a quarter on the ground! How much money does Charlie have now?

A.	58¢	**C.**	43¢
B.	33¢	**D.**	55¢

110. Sam saved up this much money:

He took 75¢ and bought candy at the candy shop. How much money does Sam have now?

A.	30¢	**C.**	100¢
B.	15¢	**D.**	25¢

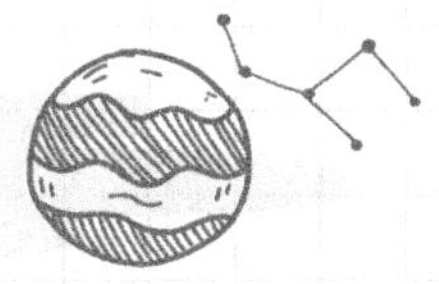

111. Donna has 2 quarters and 3 dimes. Does she have enough money to buy stickers that cost 75¢?

112. Jimmy wants to buy a new bell for his bike. It costs 99¢. He has 4 quarters. Does he have enough to buy the bell?

113. Liz went to the store and bought a chocolate chip cookie for 50¢. She gave the cashier 1 quarter and 3 dimes. How much change will Liz get back?

114. Dom earns 10¢ a day for sweeping the floor. He really wants to buy a comic book for 90¢. How many days will be have to work to save up enough money to buy the comic book?

115. Each piece of bubble gum costs 15¢. How much will it cost to buy 3 pieces of gum?

116. Chris needed to buy 10 nails to build a picture frame. He has 85¢. The nails cost 10¢ each. How much more money does Chris need?

117. Lucy went to the dollar store to buy a new pack of crayons for school. Everything in the store cost 1 dollar. Lucy has 100¢. Can she buy the crayons? Why or why not?

TIP: When counting money, count the coins in order of their value! Count the coins that are worth the most first and the ones worth the least last!

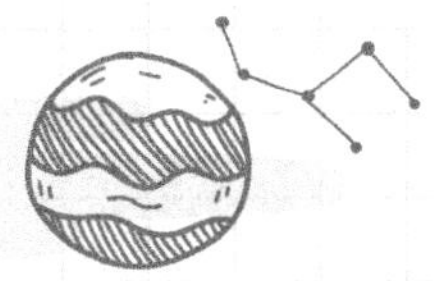

118. Parker has 68¢ and Harry has 61¢. How much more money does Parker have than Harry?

119. A bag of chips costs 25¢ at the snack stand, and a drink cost 50¢. Danny buys both and gets 25¢ back in change. How much money did Danny give the cashier?

120. Megan buys milk for 10¢, an apple for 25¢, and a hot dog for 50¢. She has 100¢ to spend on lunch. She really wants a brownie for dessert. The brownie costs 20¢. Can she get the brownie?

Look at the line plot below and answer the questions:

How long do you play outside in 1 day?

Minutes outside	5	10	15	20	25	39	35	40	45	50	55	60	65	70
	X	X	XX	XXX	X	XXXX	X	XXXXX	XXXXXX	XXXXX	X	XXXXXXX		XX

Minutes outside

121. What is the most popular amount of time for children to play outside?

122. What is the least popular amount of time for children to play outside?

123. How many minutes are 6 children outside for?

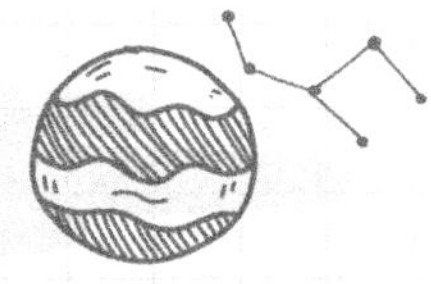

124. How many children are outside for 20 minutes a day?

125. How many children go outside for 50 minutes and 55 minutes a day?

Look at the line plot below and answer the questions:

How many stickers were sold today?

5¢ 10¢ 15¢ 20¢ 25¢ 39¢ 35¢ 40¢ 45¢ 50¢ 55¢ 60¢ 65¢ 70¢

Cost of stickers

126. How many 50¢ stickers were sold?

127. How much money was made on 5¢ stickers and 10¢ stickers combined?

128. How many more 5¢ stickers were sold than 60¢ stickers?

129. How much more money was made on 25¢ stickers than 35¢ stickers?

130. How many fewer 65¢ stickers were sold than 10¢ stickers?

TIP: If you do not understand a problem, be sure to ask a teacher or parent for help.

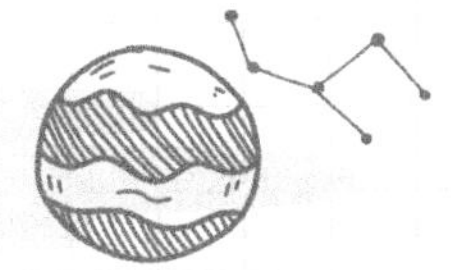

Create a line plot based on the data given. Then, answer the questions below.

What kind of pet do you have?

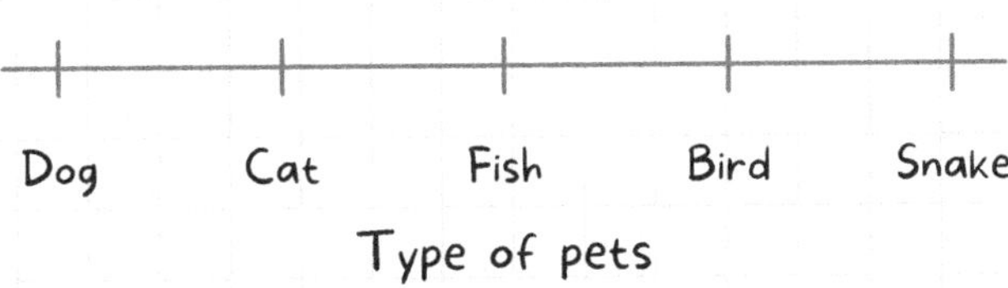

Type of pets

Dog	9
Cat	7
Fish	5
Bird	2
Snake	1

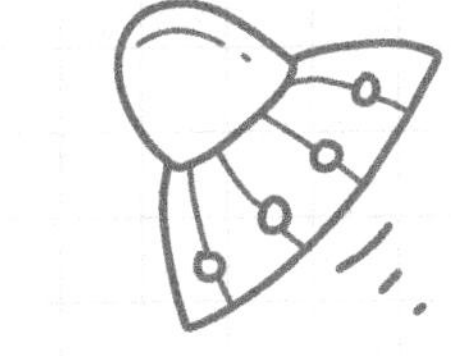

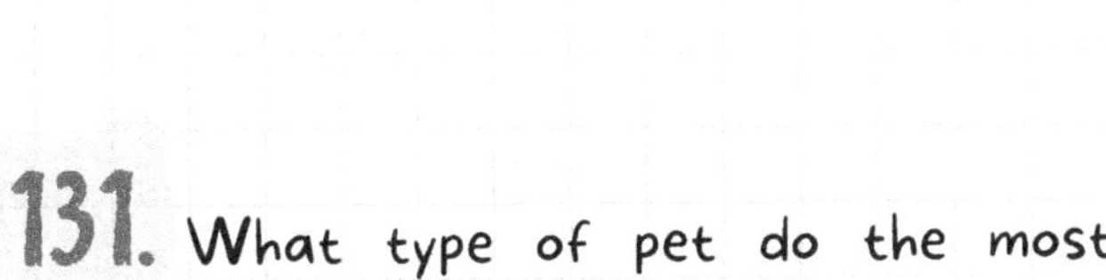

131. What type of pet do the most amount of children have?

132. What type of pet do the least amount of children have?

133. How many children have dogs and cats altogether?

134. How many more children have cats than birds?

135. How many fewer children have snakes than dogs?

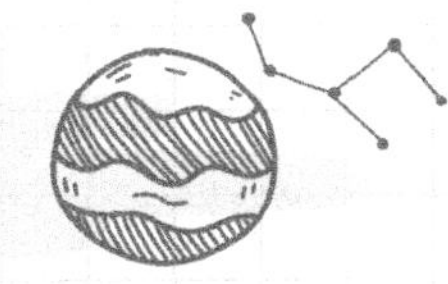

Create a line plot based on the data given. Then, answer the questions below.

How tall are you?

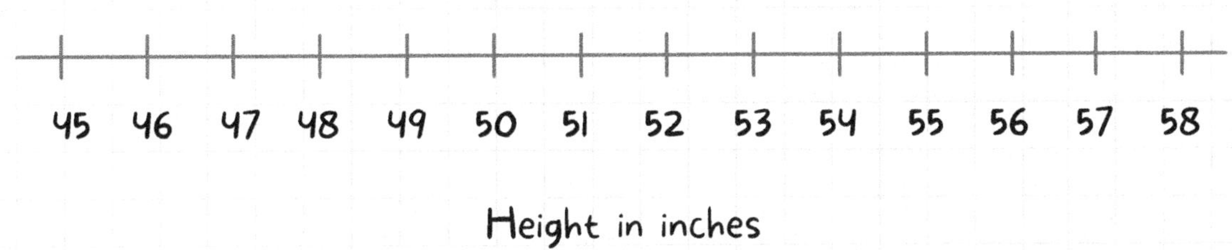

Height in inches

Height in Inches	46	47	48	49	50	51	52	53	54	55	56	58
Number of Kids	1	3	4	2	5	6	2	7	1	8	3	1

136. How many kids are 53 inches?

137. There are 8 kids that are how tall?

BRAIN HUNTER

138. How many kids are 53 inches and 55 inches altogether?

139. How many kids are shorter than 51 inches?

140. How many kids are taller than 53 inches?

TIP: When looking at a line plot, remember that some choices may have zero. If a choice has zero, that would be the fewest on the line plot.

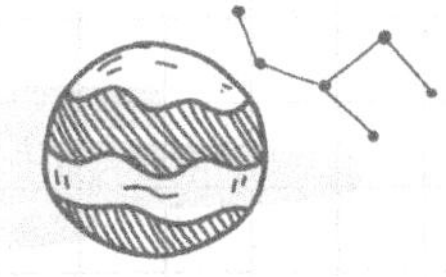

Look at the bar graph below and answer the questions:

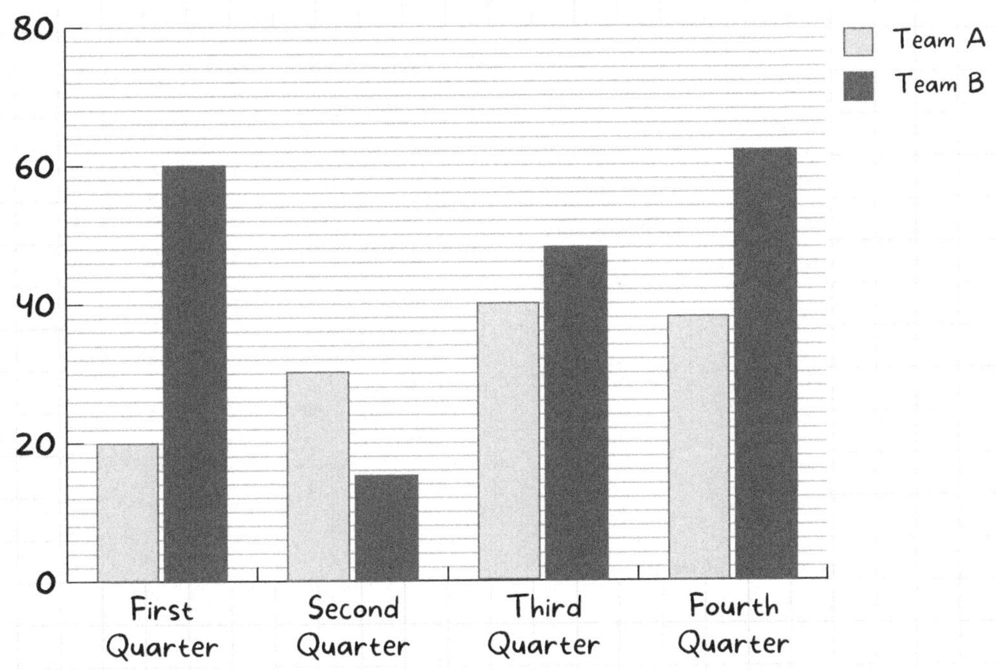

141. How many more points did Team B score in the first quarter than Team A?

142. How many fewer points did Team A score in the Fourth Quarter than Team B?

143. How many points did Team A score in the third and fourth quarter combined?

144. How many points were scored by both teams in the second quarter in all?

145. What was the final score? Who won the whole game?

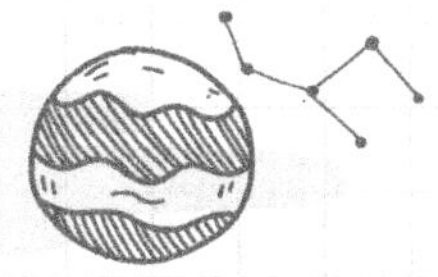

Look at the bar picture graph below and answer the questions:

Favorite Sports

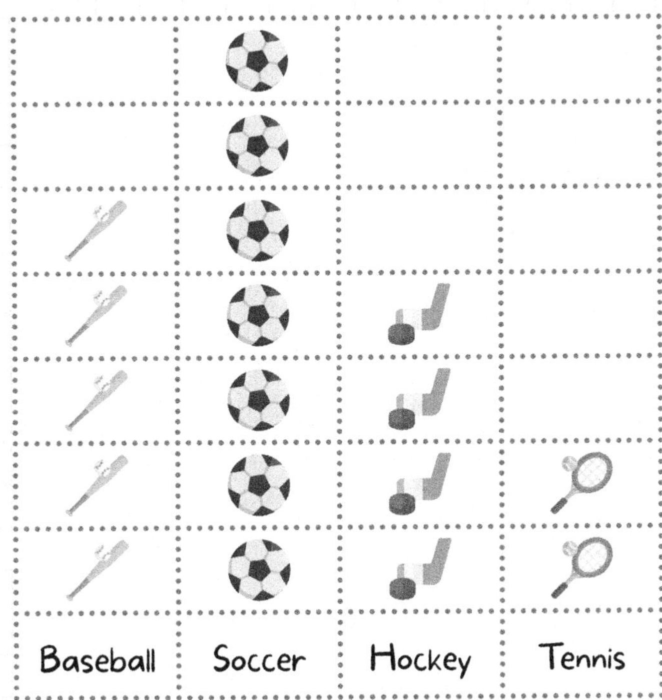

146. What sport did most kids like best?

147. What was the least favorite sport?

148. How many kids liked baseball and soccer in total?

149. If 2 more kids picked tennis, how many kids would have picked it?

150. How many fewer kids liked hockey than soccer?

TIP: When determining how many are less or more than a given number, do not count the given amount in your total. Go one above or one below.

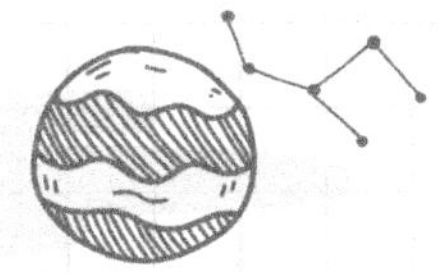

Create a picture graph based on the following data. Answer the questions below.

What did you have for lunch?

Lunch options	Number of People
Pizza	3
Hot Dog	5
Salad	7
Sandwich	10
Soup	4

151. How many fewer people picked soup than sandwich?

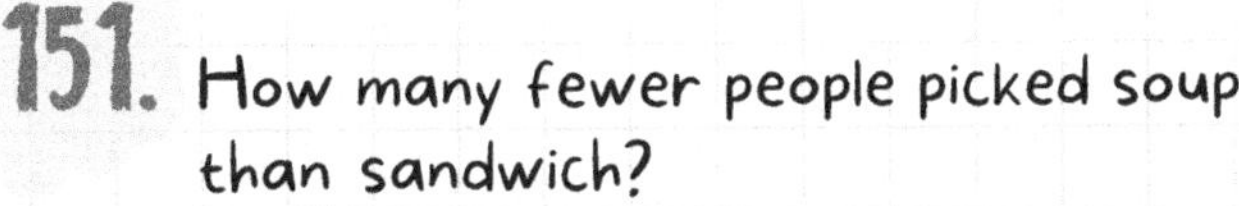

152. How many people picked sandwich and salad altogether?

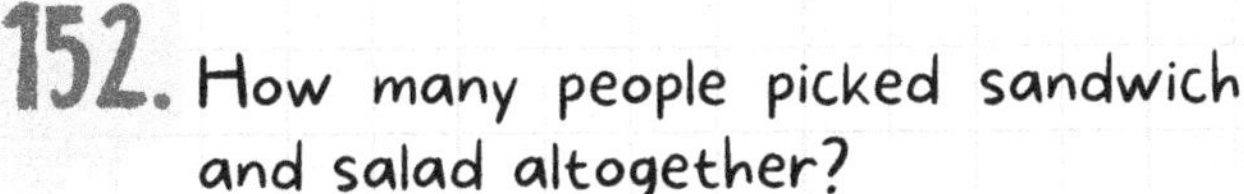

153. What lunch did the most people have?

154. What lunch did the fewest people have?

155. How many more people would have needed to eat pizza for it to have been equal to sandwich?

Create a bar graph based on the following data. Answer the questions below.

How many minutes a day do you read?

Minutes	Number of Kids
10	5
20	12
30	15
45	4
60	2

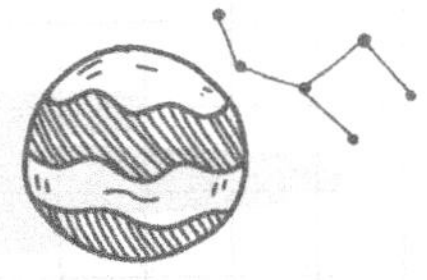

156. How many kids read for 45 minutes a day?

SHOW YOUR WORK

157. How many more kids read for 30 minutes than 20 minutes?

SHOW YOUR WORK

158. What is the most popular amount of time to read in a day?

SHOW YOUR WORK

159. How many kids read for less than 30 minutes a day?

SHOW YOUR WORK

160. How many kids read for more than 20 minutes a day?

SHOW YOUR WORK

CHAPTER 4

GRADE 2

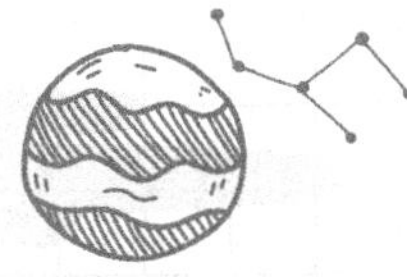

1. Circle the polygons below:

2. Circle the polygons below:

3. Circle the polygons below:

4. Circle the polygons below:

5. Find the polygon that has 3 sides and 3 vertices.

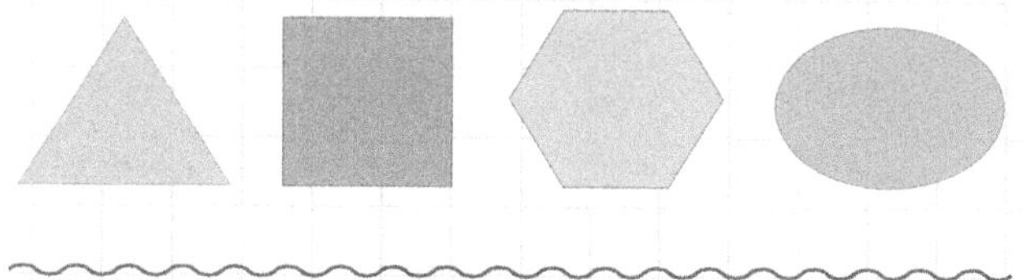

6. Find the polygon that has 4 sides and 4 vertices.

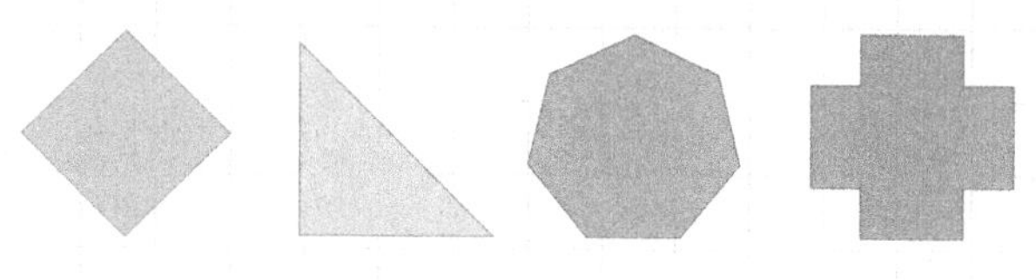

TIP: Vertices are another word for corners. Two flat sides come together at a vertex.

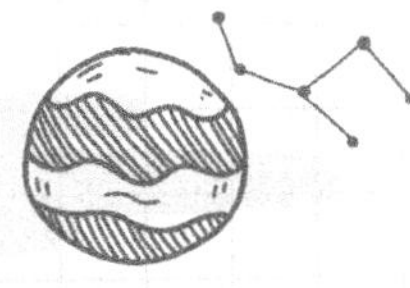

7. Find the polygon that has 8 sides and 8 vertices.

8. Find the polygon that has 5 sides and 5 vertices.

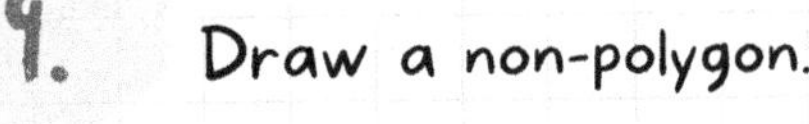

9. Draw a non-polygon.

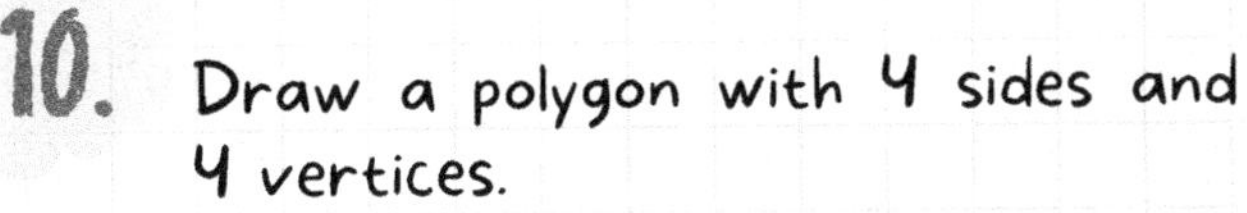

10. Draw a polygon with 4 sides and 4 vertices.

11. Draw a non-polygon with 5 sides.

12. Draw a polygon with 5 sides and 5 vertices.

13. Draw an open non-polygon with a curve.

14. Draw lines to divide the rectangle into square units.

15. How many square units are in the rectangle above?

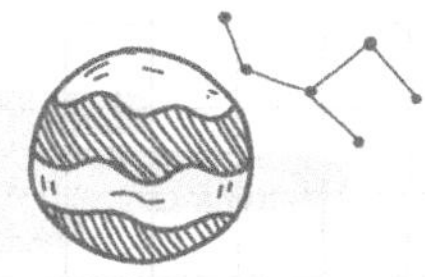

16. Draw lines to divide the rectangle into square units.

17. How many square units are in the rectangle above?

18. Draw lines to divide the rectangle into squares.

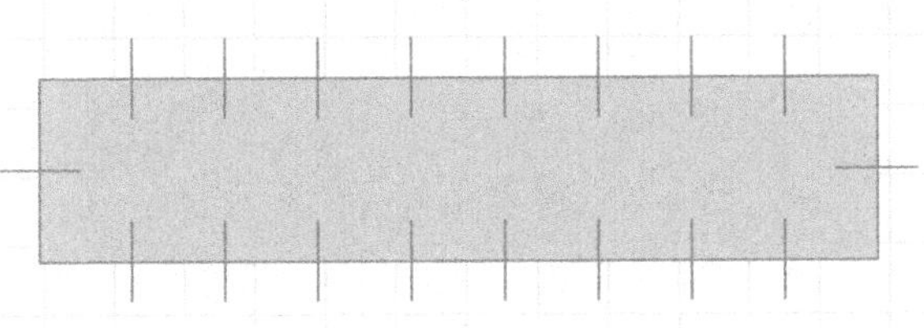

19. How many square units are in the rectangle above?

20. Divide this rectangle into 4 equal columns.

21. Divide this rectangle into 5 equal rows.

TIP: When you divide a shape make sure you count how many rows or columns you created. The number of divisions does not equal the number of columns or rows. For example, if you make 3 divisions, you created 4 columns.

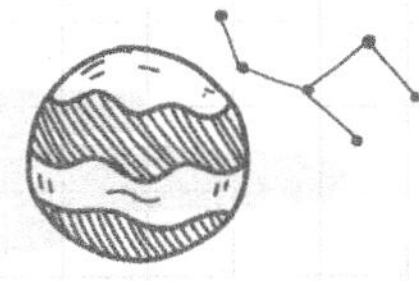

22. Divide this rectangle into 3 equal columns.

23. Divide this rectangle into 6 equal rows.

24. Divide this rectangle into 3 rows and 4 columns.

25. How many square units are in the rectangle above?

26. Divide this rectangle into 2 rows and 5 columns.

27. How many square units are in the rectangle above?

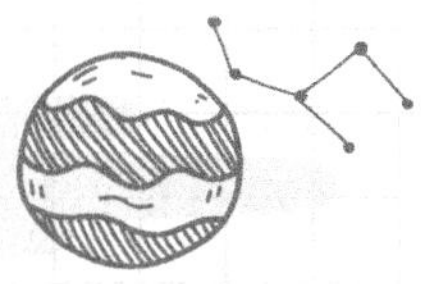

28. Divide these circles in half in 2 different ways.

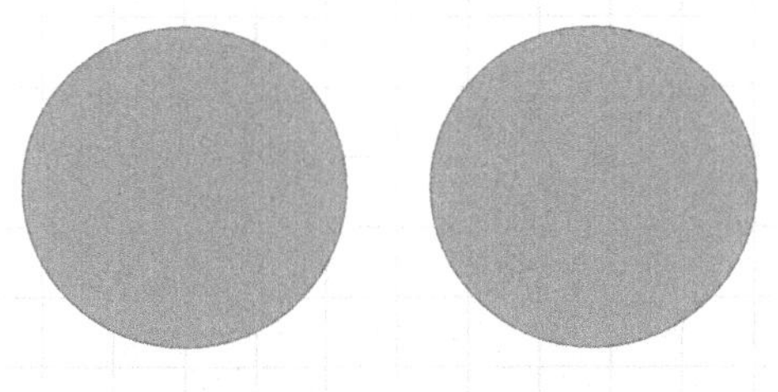

29. Divide these rectangles in half in 2 different ways.

30. Divide these circles into fourths 2 different ways.

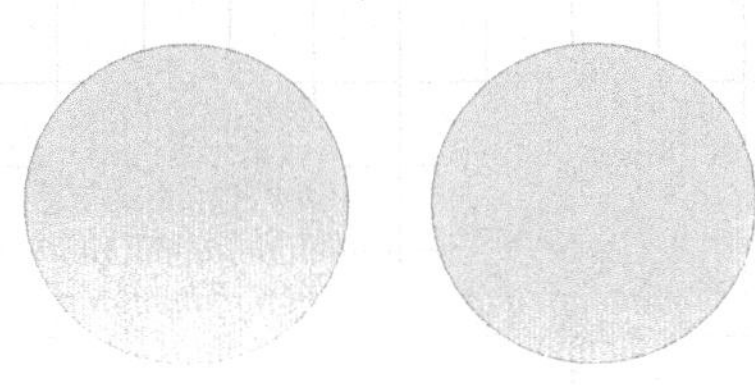

31. Divide these squares into fourths 2 different ways.

32. Which shape is divided into fourths?

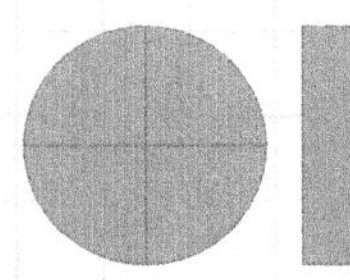  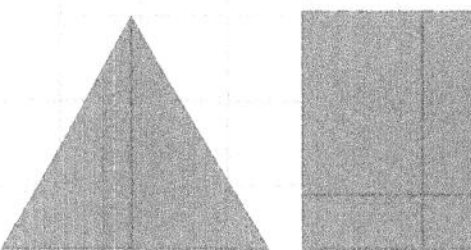

33. Which shape is divided in half?

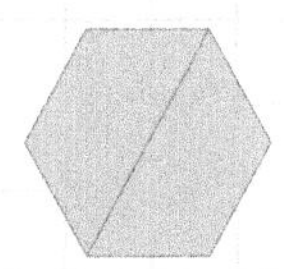 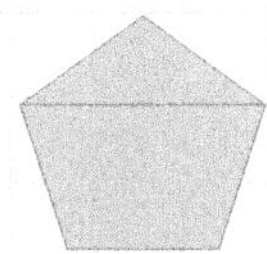 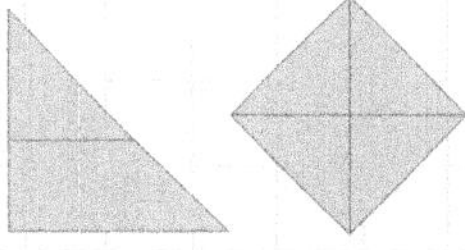

TIP: When dividing a shape into fractions, all parts need to be the same size or equal.

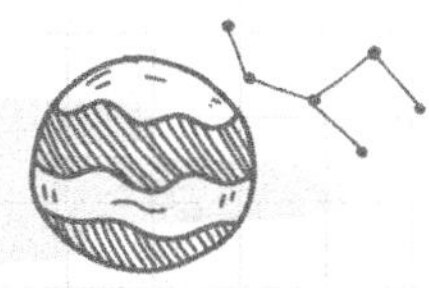

34. Which shape is divided into thirds?

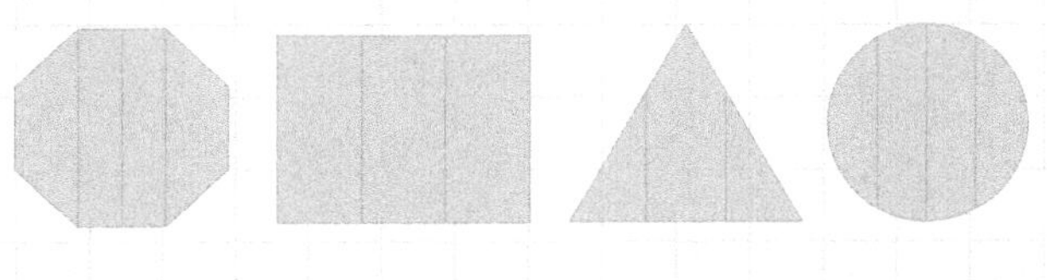

35. How is this shape divided?

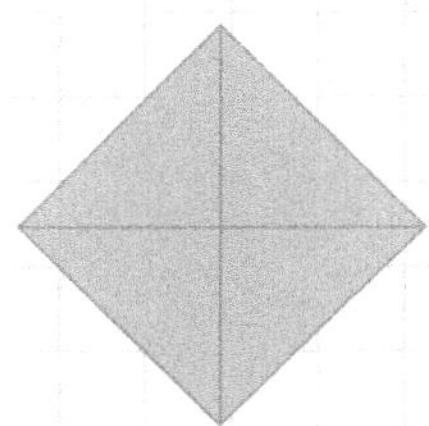

A. The shape is divided into 4 fourths.

B. The shape is divided into 3 thirds.

C. The shape is divided into 2 halves.

D. The shape is divided into 1 fourth.

36. How is this shape divided?

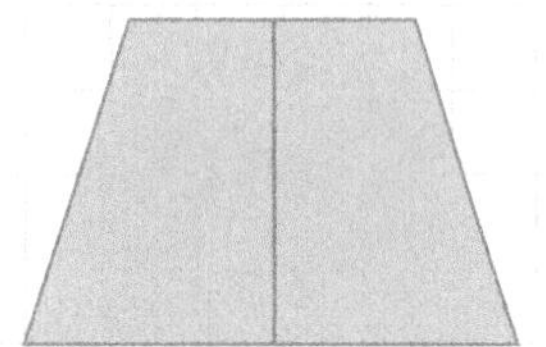

A. The shape is divided into 3 thirds.
B. The shape is divided into 2 halves.
C. The shape is divided into 4 fourths.
D. The shape is divided into 5 fifths.

37. How is this shape divided?

A. The shape is divided into 5 fifths.
B. The shape is divided into 2 halves.
C. The shape is divided into 4 fourths.
D. The shape is divided into 3 thirds.

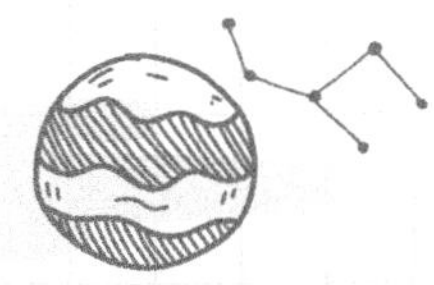

38. Color half of the triangle.

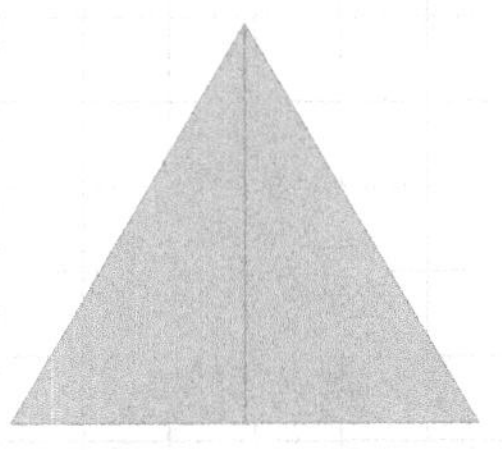

39. Color a third of the rectangle.

40. Color a quarter of the circle.

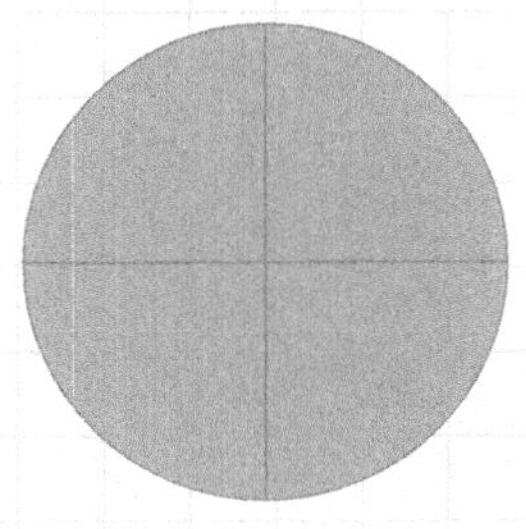

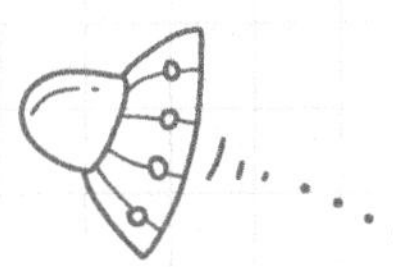

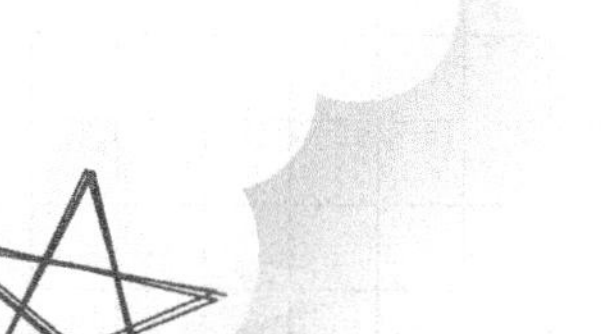

CHAPTER 5

GRADE 2

5 CHAPTER

ASSESSMENT

SECTION: ASSESSMENT

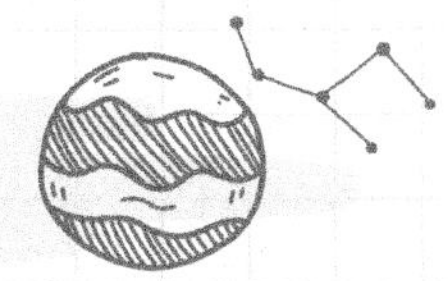

1. What addition equation represents joining these two groups?

A. 10 + 9 = 19
B. 9 + 9 = 18
C. 10 + 8 = 18
D. 9 + 8 = 17

2. Draw a picture to find the missing number: ? + 28 = 82

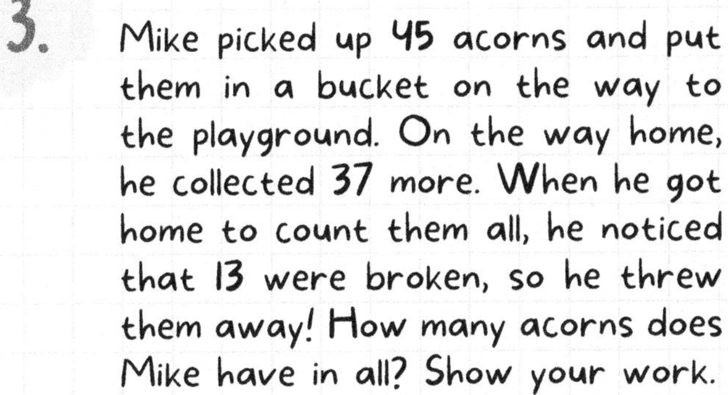

3. Mike picked up 45 acorns and put them in a bucket on the way to the playground. On the way home, he collected 37 more. When he got home to count them all, he noticed that 13 were broken, so he threw them away! How many acorns does Mike have in all? Show your work.

4. Write a subtraction equation to match this picture.

5. My basketball team scored 64 points. The opposing team scored 59 points. How many points did we win by? Show your work!

6. Which double fact would help you solve the problem 8 + 9 = 17?

A. 5 + 5 = 10
B. 6 + 6 = 12
C. 7 + 7 = 14
D. 9 + 9 = 18

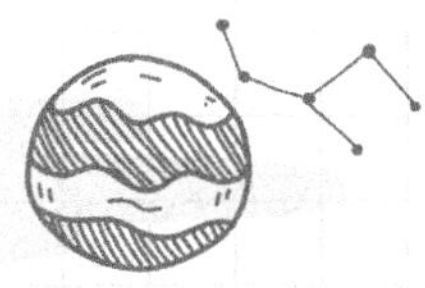

7. If you know that 10 + 2 = 12, then it should help you solve which problem?

A. 9 + 3 = 12
B. 10 + 10 = 20
C. 9 + 10 = 19
D. 6 + 6 = 12

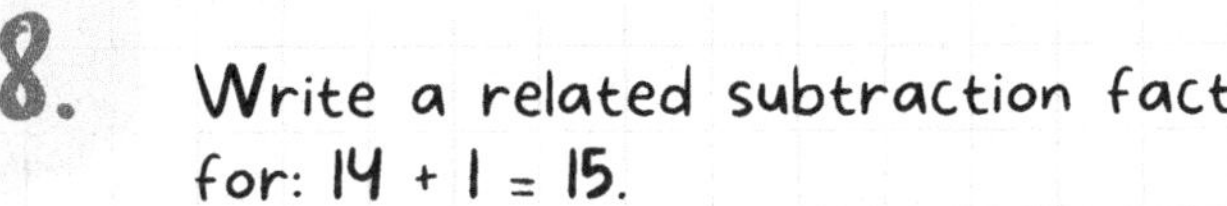

8. Write a related subtraction fact for: 14 + 1 = 15.

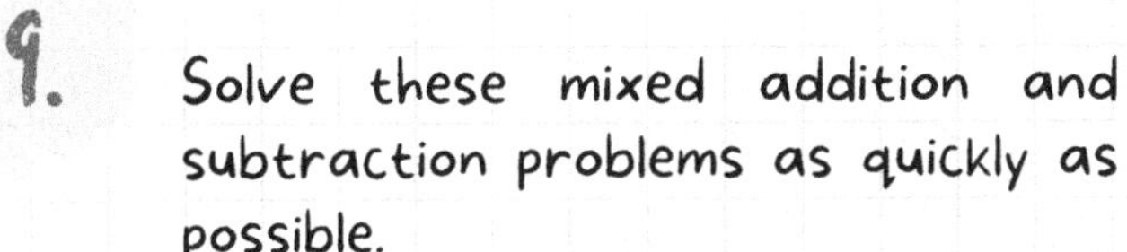

9. Solve these mixed addition and subtraction problems as quickly as possible.

9 + 10 = __	13 + 5 = __	17 - 12 = __
4 + 12 = __	6 + 6 = __	15 - 6 = __
19 - 8 = __	12 - 10 = __	1 + 14 = __

10. Solve these addition equations with a missing addend.

12 + __ = 19	__ + 11 = 17	2 + __ = 17
__ + 6 = 14	5 + __ = 10	__ + 9 = 15

11. How many watermelons are there? Circle odd or even.

______ watermelon Odd Even

TIP: When solving an addition equation with a missing addend, you can either use a related subtraction fact or start at the known number and count on until you reach the sum.

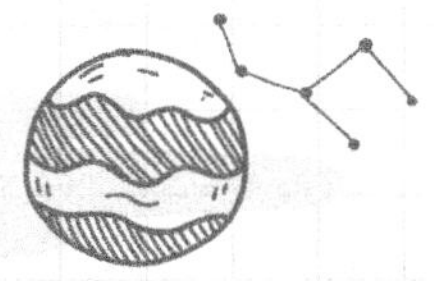

12. Write an addition equation with 2 equal addends to match the picture:

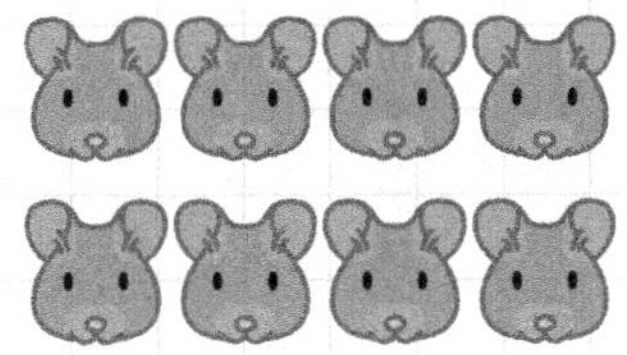

13. There are 3 dogs. How many feet in all? Draw a picture to show your work!

14. Write an addition sentence with equal addends to match this array:

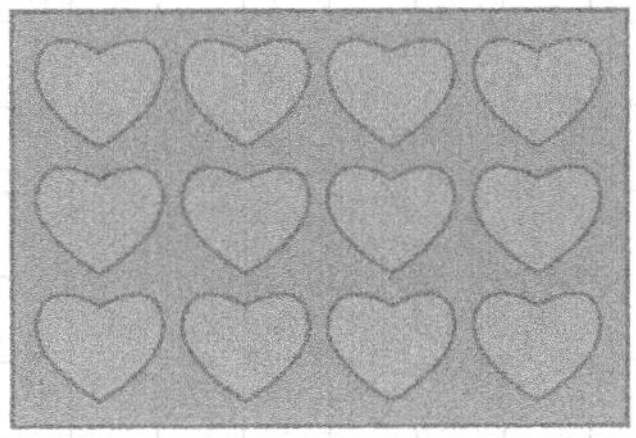

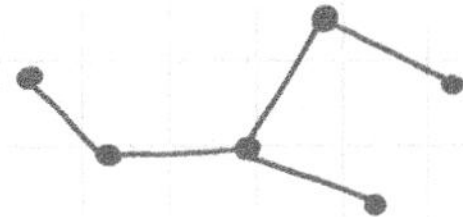

15. Draw an array to match the addition sentence: 3 + 3 + 3 + 3 + + 3 = 15

16. How many tens are in the number 741?

17. Which number has 5 hundreds, 2 tens, and 9 ones?

A. 295
B. 529
C. 952
D. 292

18. Start at 245 and count on by 5s.

245, ___ , ___ , ___ , ___ , ___ , ___ , ___

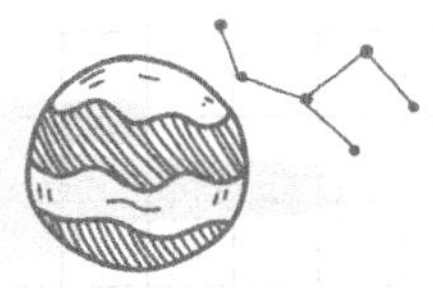

19. IWrite the number 126 in expanded form.

20. Compare these two numbers using >, <, and =.

754 ____ 544

21. Solve these addition equations.

62 + 35	27 + 35	55 + 40
38 + 24	64 + 26	39 + 17

22. Solve these subtraction equations.

32 - 17	86 - 38	78 - 25
91 - 55	46 - 25	67 - 43

23. Solve this addition equation. Show your work!

17 + 21 + 5 + 46 =

24. Write an equation that represents 100 more than 33.

TIP: When adding an equation with multiple addends, add all the ones first, then all the tens. Last, add the tens and ones together.

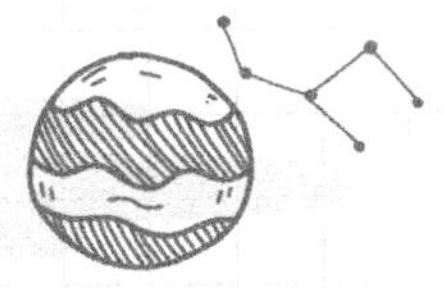

25. Donald's favorite author was at the bookstore signing her new book. The first 125 people in line got a free book. Donald got there early and was 76th in line. How many people in line after Donald also got a free book?

26. What would be the most efficient tool to measure the height of the refrigerator?

A. A measuring cup
B. A ruler
C. A measuring tape
D. Thermometer

27. Measure the caterpillar to the closest centimeter.

28. The water fountain measures 24 inches or 18 paper clips. How many fewer paper clips than inches are needed to measure the water fountain?

29. Christie's bathroom is 5 yards wide. When she measured it with a ruler, she measured 15 feet. Why did Christie get different measurements when measuring the width of her bathroom?

A. Feet and yards are equal.
B. Yards and smaller than feet.
C. Feet are bigger than yards.
D. Yards are bigger than feet.

30. Estimate how tall a flagpole may be.

A. 2 yards
B. 24 inches
C. 20 feet
D. 200 centimeters

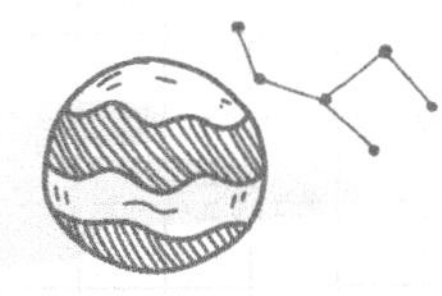

31. An elephant is about 132 inches tall. A giraffe is about 180 inches tall. How much taller is a giraffe compared an elephant?

32. Draw a line that is 6 inches long in all. 3 of the inches should be red, and the rest should be blue. Write an equation to show the total.

33. Draw a picture to solve. 1 bat is 3 feet long. How long would it be if you lined up 5 bats?

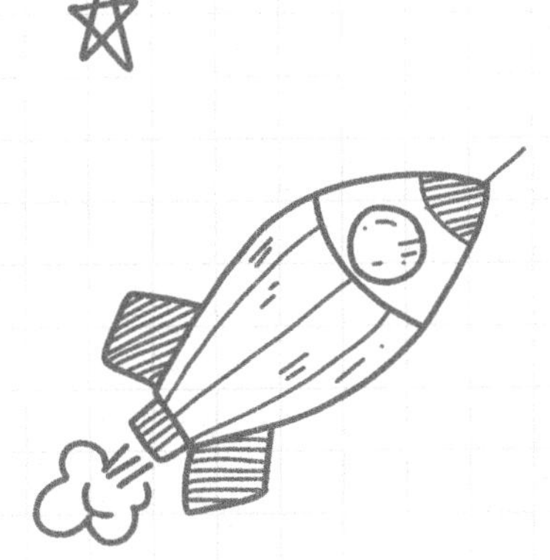

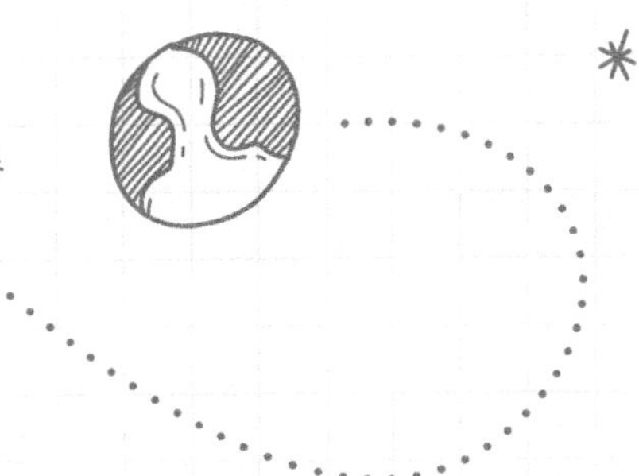

34. Place the number **267** on the number line below.

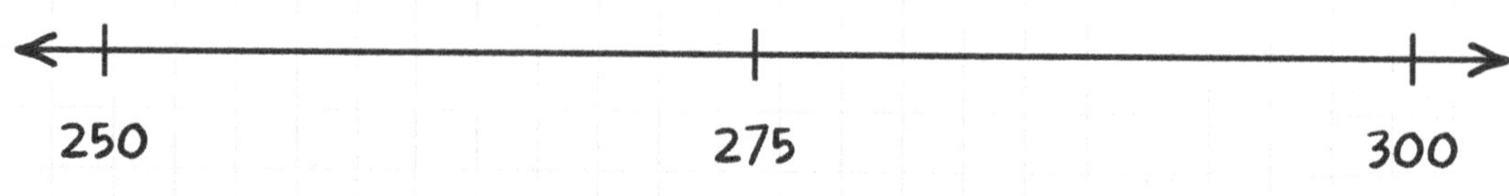

35. Show how to solve this equation on the number line: **521 - 10** = ?

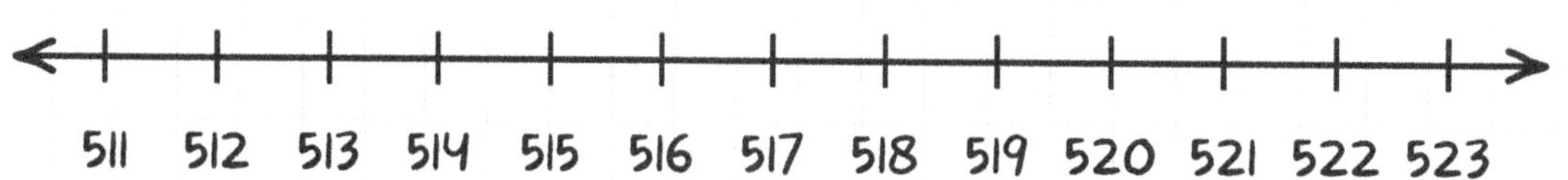

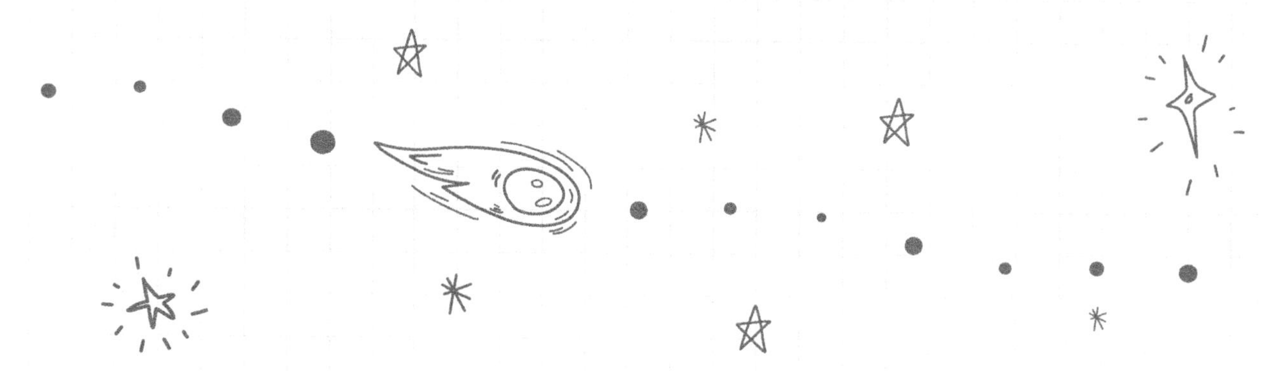

TIP: When counting on a number line you don't have to count by ones. You can count by 2s, 5s or 10s to count faster!

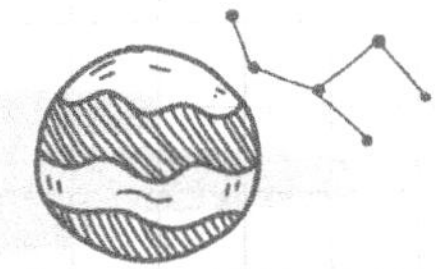

36. What time does the clock show?

A. 10:05
B. 10:25
C. 5:10
D. 10:30

37. Show 4:45 on the clock below:

38. What times does this clock show:

01 : 15

A. Quarter past 1:00
B. Half past 1:00
C. Half past 5:00
D. Quarter of 2:00

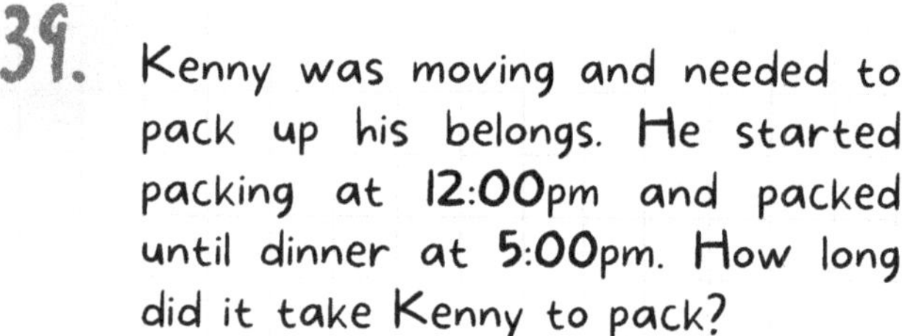

39. Kenny was moving and needed to pack up his belongs. He started packing at 12:00pm and packed until dinner at 5:00pm. How long did it take Kenny to pack?

A. 2 hours
B. 8 hours
C. 6 hours
D. 5 hours

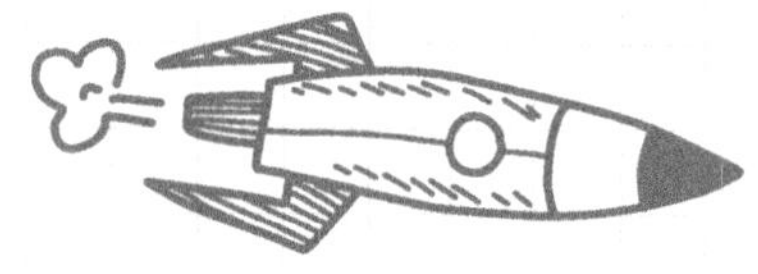

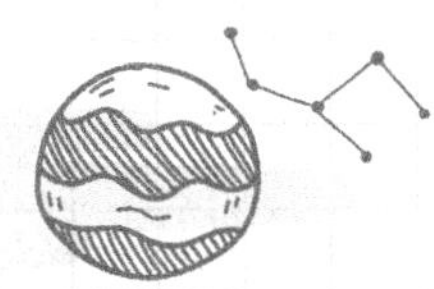

40. What times does this clock show:

12 : 30

A. Quarter past 2:00
B. Half past 12:00
C. Quarter of 2
D. 5 of 12:00

41. Count the money below.

A. 75¢
B. 69¢
C. 70¢
D. 65¢

42. Count the money below.

A. 55¢
B. 46¢
C. 61¢
D. 56¢

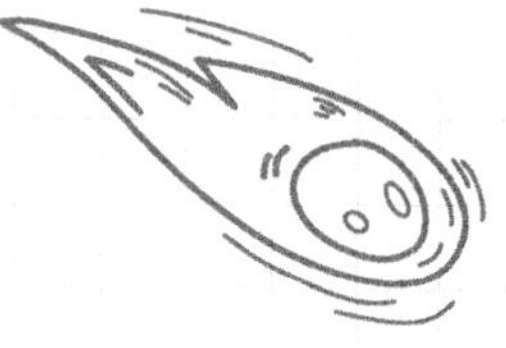

TIP: You can combine and add like coins together first. Then add all the amounts together.

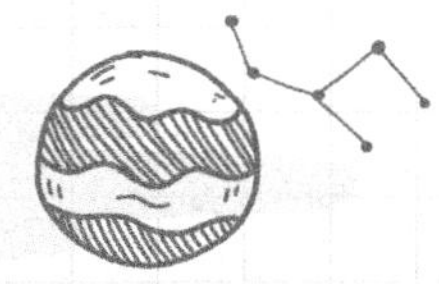

43. Erin saved up this much money:

She needed to pay her brother back the **50¢** that he let her borrow last week at the snack stand. How much money does Erin have now?

A. 28¢
B. 78¢
C. 25¢
D. 20¢

44. Tony went to the store and bought an apple for **35¢** and a water for **50¢**. He gave the cashier 4 quarters. How much change will Tony get back?

45. At the school store, pencils cost **20¢** each. Kelly has **80¢**. How many pencils can she buy?

46. Circle the polygons below:

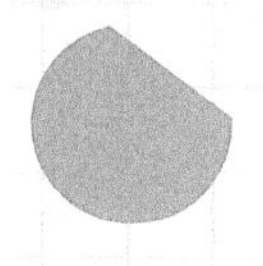   

47. Find the polygon that has 6 sides and 6 vertices.

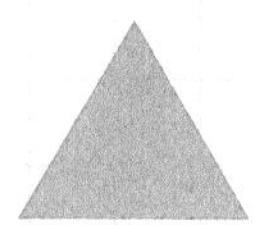   

48. Draw a non-polygon with 3 straight sides.

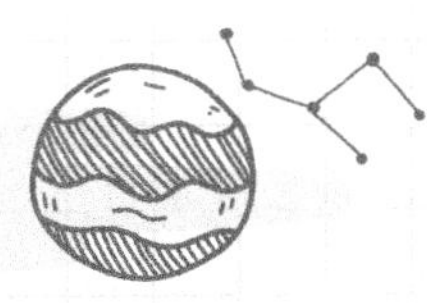

49. Divide this square into 4 rows and 4 columns.

50. How many square units are in the rectangle above?

ANSWERS AND EXPLANATIONS

GRADE 2

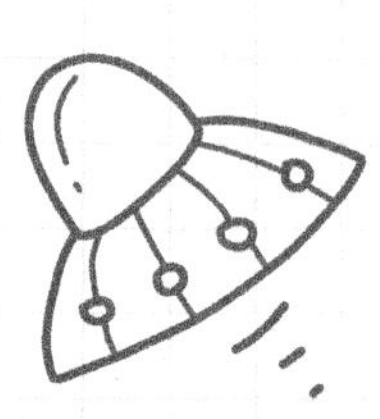

ANSWER AND EXPLANATION

SECTION 1: REPRESENT ADDITION AND SUBTRACTION

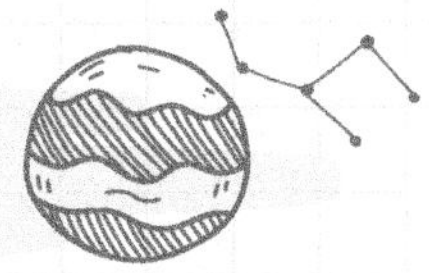

1. B.
The first group has 10, and the second group has 7. When you join them together and count them in all, the sum is 17. The corresponding addition equation would be (B) 10 + 7 = 17.

2. A.
The first group has 15, and the second group has 12. When you join them together and count them in all, the sum is 27. The corresponding addition equation would be (A) 15 + 12 = 27.

3. The first group has 18 stars, and the second group has 10 stars. When you join them together and count them in all, the sum is 28. The corresponding addition equation would be 18 + 10 = 28.

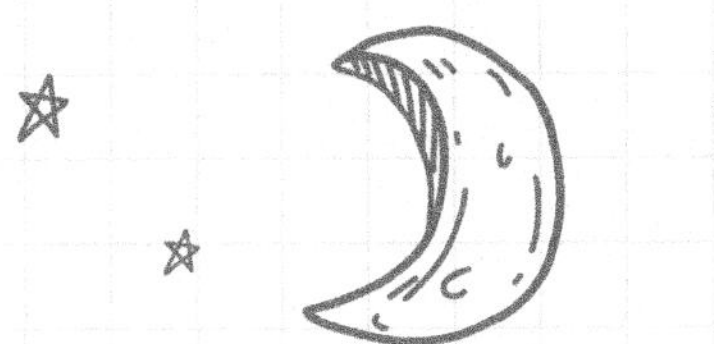

4. Any visual representation that shows a group of 12 and a group of 9 is correct. There should clearly be 2 separate groups. An example would be:

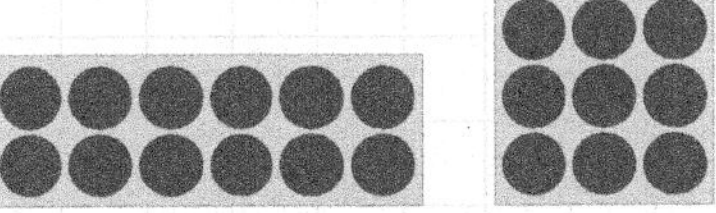

5. Any visual representation that shows a group of 19 and a group of 6 is correct. There should clearly be 2 seperate groups. An example would be:

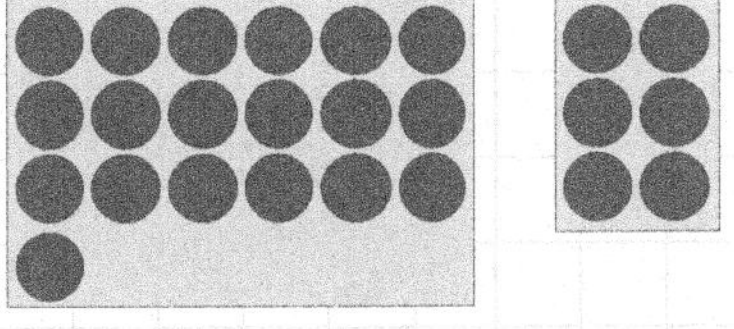

6. D.
The first group has 33 blocks, and the second group has 47 blocks. Add the ones: 7 + 3 = 10, and then add the tens: 30 + 40 = 70. When you join them together and count them in all, the sum is 80. The corresponding addition equation would be (D) 33 + 47 = 80.

7. C.
The first group has 65 blocks, and the second group has 26 blocks. Add the ones: 5 + 6 = 11, and then add the tens: 60 + 20 = 80. When you join them together and count them in all, the sum is 91. The corresponding addition equation would be (C) 65 + 26 = 91.

8. D.
The first group has 18, the second group has 10, and the third group has 9. When you join them together and count them in all, the sum is 37. The corresponding addition equation would be (D) 18 + 10 + 9 = 37.

9. C.
The first group has 21, the second group has 16, and the third group has 4. When you join them together and count them in all, the sum is 41. The corresponding addition equation would be (C) 21 + 16 + 4 = 41.

10. Any visual representation that shows a group of 17, a group of 15, and a group of 2 is correct. There should clearly be 3 separate groups.

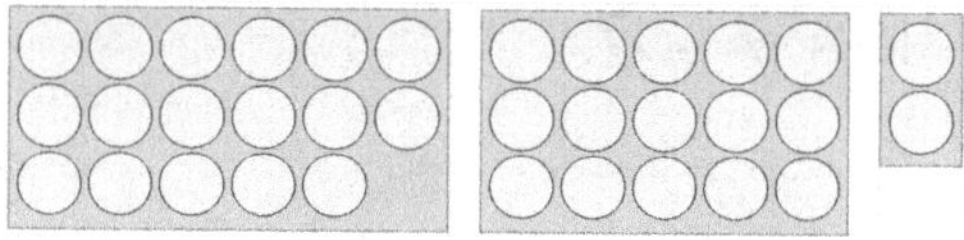

11. Any visual representation that shows a group of 31, a group of 9, and a group of 10 is correct. There should clearly be 3 separate groups.

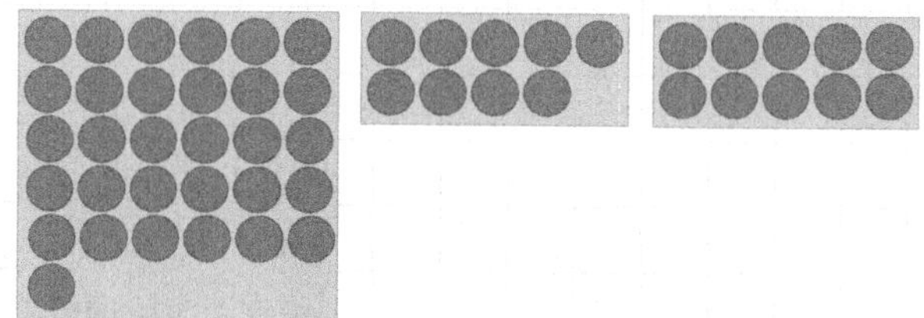

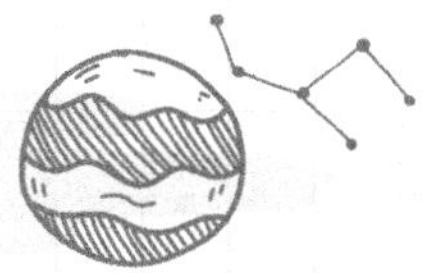

12. Any visual representation that depicts a strategy to show that the missing number is 23 is correct. Students can visually show 65 - 42 = 23, or students can draw 42 objects and count on 23 more until they reach 65. Students may draw representation in base 10 rods and units.

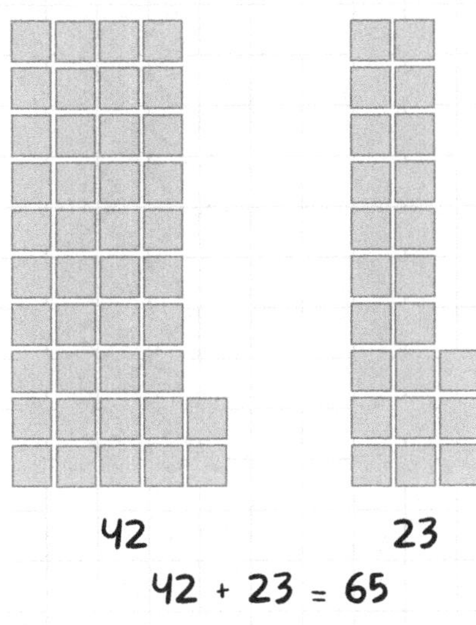

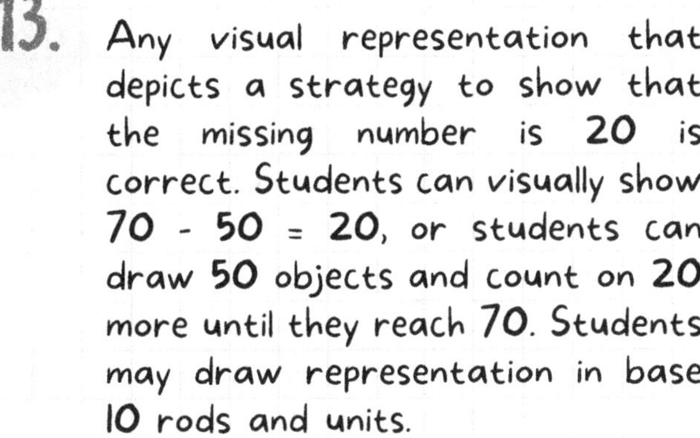

13. Any visual representation that depicts a strategy to show that the missing number is 20 is correct. Students can visually show 70 - 50 = 20, or students can draw 50 objects and count on 20 more until they reach 70. Students may draw representation in base 10 rods and units.

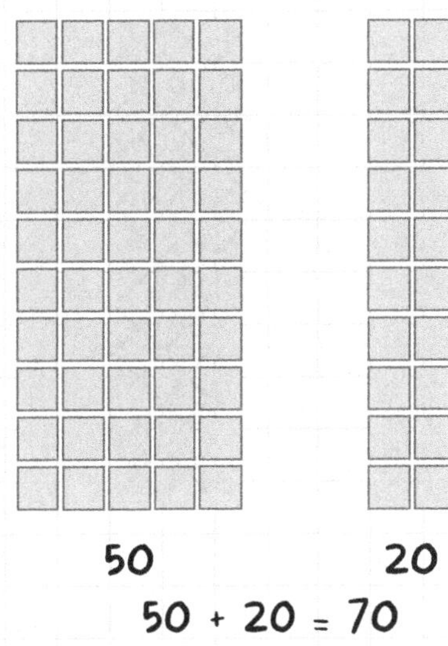

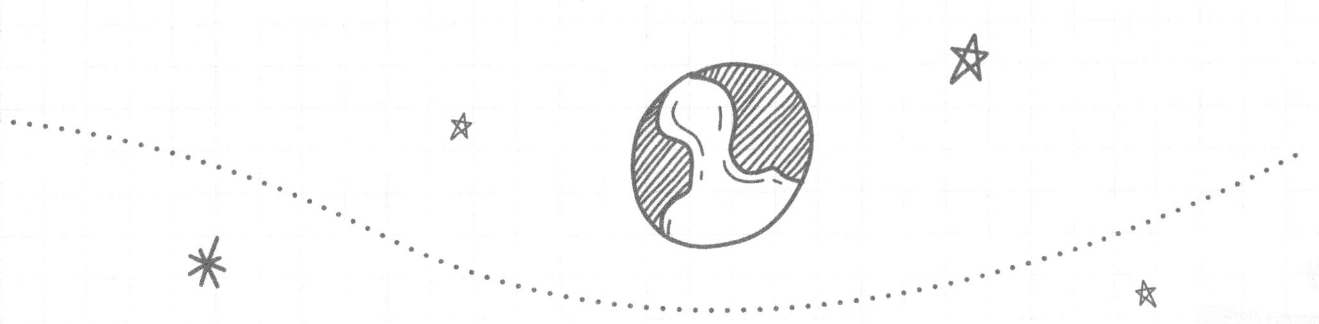

14. Any visual representation that depicts a strategy to show that the missing number is **25** is correct. Students can visually show **50 - 25 = 25**, or students can draw **25** objects and count on **25** more until they reach **50**. Students may draw representation in base 10 rods and units.

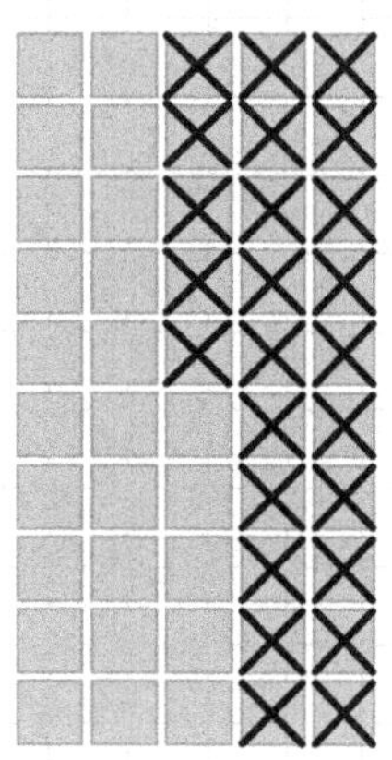

50

50 - 25 = 25

15. Any visual representation that depicts a strategy to show that the missing number is **55** is correct. Students can visually show **85 - 30 = 55**, or students can draw **30** objects and count on **55** more until they reach **85**. Students may draw representation in base 10 rods and units.

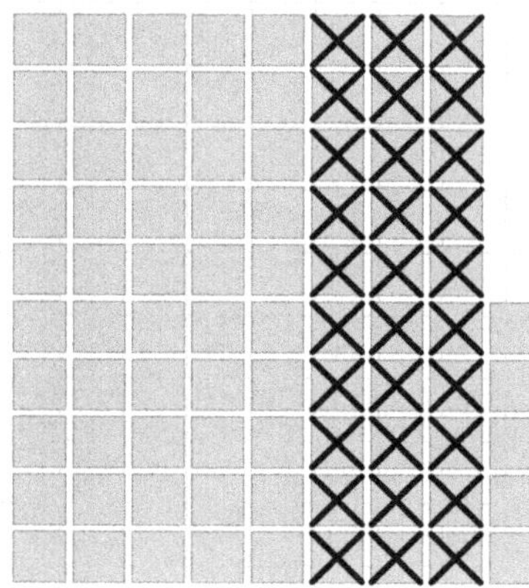

85

85 - 30 = 55

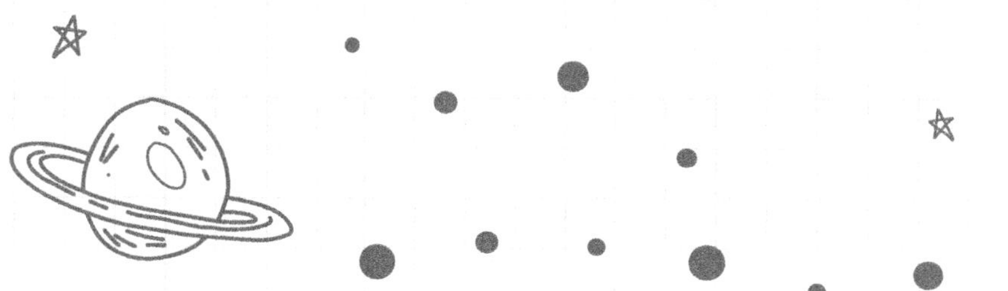

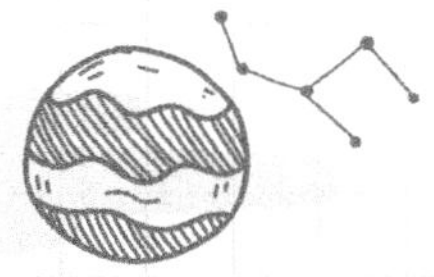

16. Since this problem is asking how many there are in all, students should join the two groups and write an addition equation. There are 35 in the first group and 62 in the second group. The equation should be 35 + 62. To solve, add the ones (5 + 2 = 7) and add the tens (30 + 60 = 90). The sum is 97. 35 + 62 = 97 stickers.

17. Students should start with Jake's 47 rocks and count on until they reach 100. They would need to count on 53 more. Another way to solve would be to take the total of 100 rocks and subtract Jake's 47 rocks. This would leave a difference of 53, which would be the amount that his sister collected.

18. Start with Levi's 32 cents and then count on until they reach 50. They would count on 18 more. Another way to solve would be to take the total of 50 cents and subtract Levi's 32 cents. This leaves a difference of 18, which would be the amount that his mom gave him.

19. This is a two step problem. Step one is to combine the two groups of blueberries. There are 43 in the first group and 42 in the second group. The equation is 43 + 42. To solve, students will add the ones (3 + 2 = 5) and then add the tens (40 + 40 = 80). The sum is 85.

The second step is to take 10 away from the total. The equation is 85 - 10. To solve, subtract the ones (5 - 0 = 5) and then subtract the tens 80 - 10 = 70. The difference is 75. (85 - 10 = = 75) blueberries.

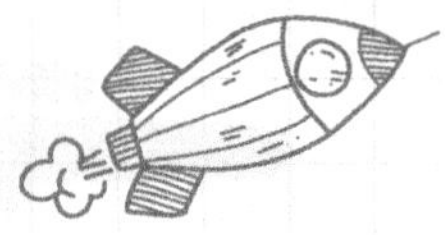

20. This is a two step problem. The first step is to combine the two groups of cups. There are 60 in the first group and 33 in the second group. The equation is 60 + 33. To solve, students will add the ones (0 + 3 = 3) and then add the tens (60 + 30 = 90). The sum is 93.

The second step is to take 73 away from the total. The equation is 93 - 73. To solve, students will subtract the ones (3 - 3 = 0) and then subtract the tens (90 - 70 = = 20). The difference is 20. 93 - 73 = 20 cups.

21. B.
The group of hearts has 22 total. There are 11 hearts that are crossed out. There are 11 hearts that remain. The corresponding subtraction equation would be (B) 22 - 11 = 11.

22. A.
The group of stars has 32 total. 14 stars are crossed out. 18 stars remain. The corresponding subtraction equation would be (A) 32 - 14 = 18.

23. The group of smileys has 24 total. 18 smileys are crossed out. 6 smileys remain. The corresponding subtraction equation would be 24 - 18 = 6.

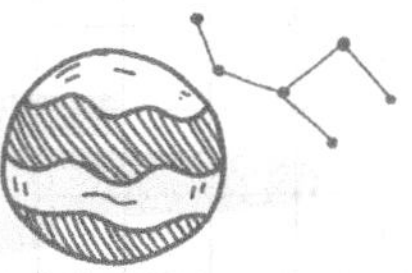

24. Any visual representation that shows a large group of 45 with 22 objects crossed out is correct. 23 objects should remain. Students may draw pictures in base 10 rods and units.

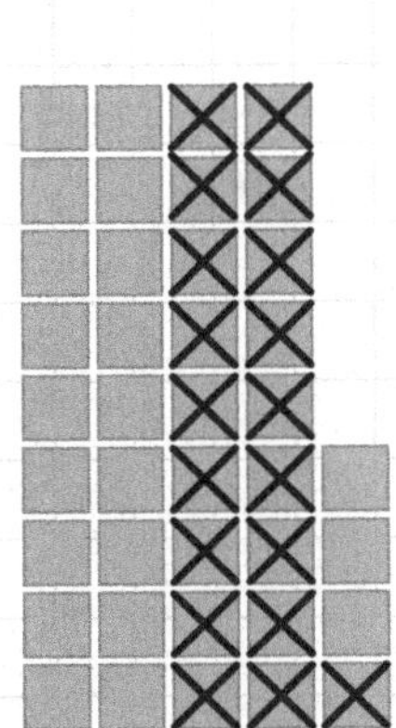

45

45 - 22 = 23

25. Any visual representation that shows a large group of 31 with 20 objects crossed out is correct. 11 objects should remain. Students may draw pictures in base 10 rods and units.

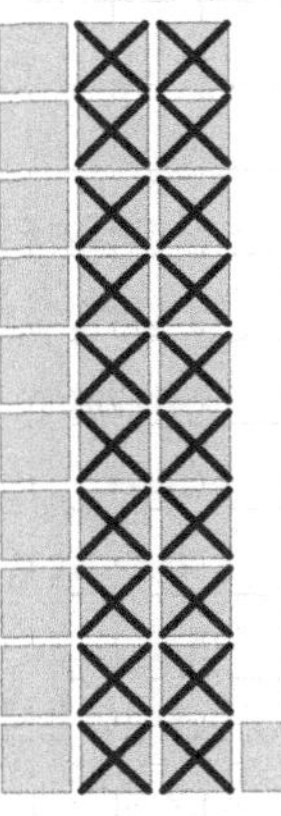

31

31 - 20 = 11

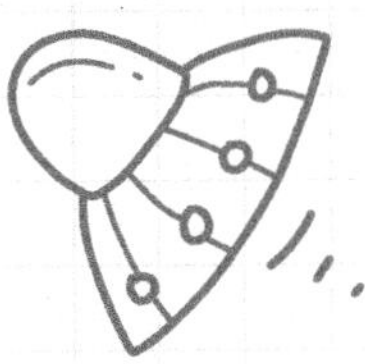

26. D.

The group of blocks has 53 total. 15 blocks are crossed out. 38 blocks remaining. The corresponding subtraction equation would be (D) 53 - 15 = 38.

27. C.
The group of blocks has 84 total. 38 blocks are crossed out. 46 blocks remain. The corresponding subtraction equation would be (C) 84 - 38 = 46.

28. Any visual representation that depicts a strategy to show that the missing number is 17 is correct. Students can visually show 54 - 37 = 17, or students can draw 37 objects and count on 17 more until they reach 54. Students may draw representation in base 10 rods and units.

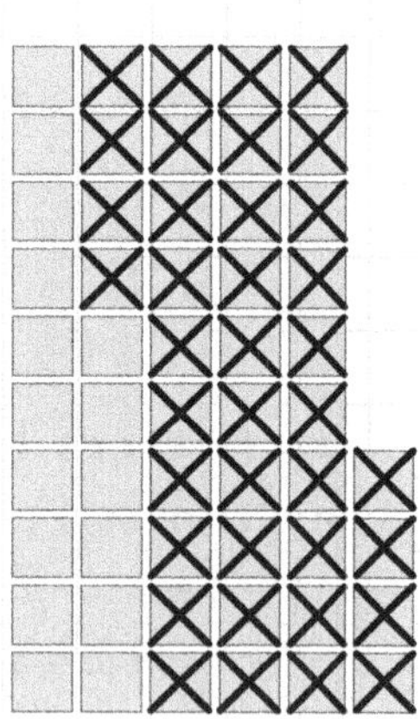

54
54 - 37 = 17

29. Any visual representation that depicts a strategy to show that the missing number is 74 is correct. Students can visually show 95 - 21 = 74, or students can draw 21 objects and count on 74 more until they reach 95. Students may draw representation in base 10 rods and units.

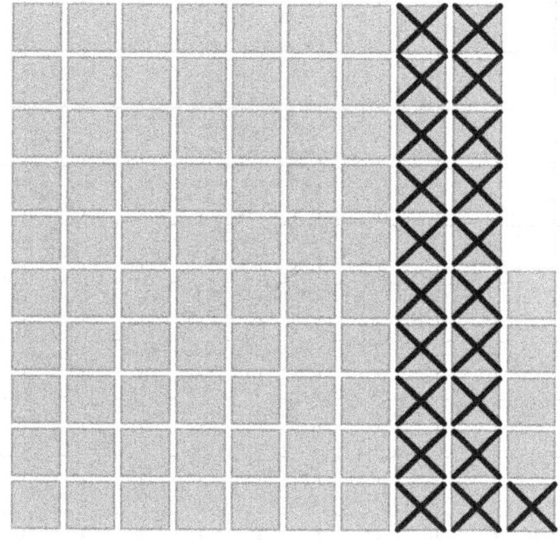

95
95 - 21 = 74

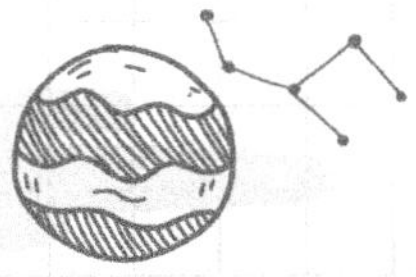

30. Any visual representation that depicts a strategy to show that the missing number is 77 is correct. Students can visually show 49 + 28 = 77. Students may draw representation in base 10 rods and units.

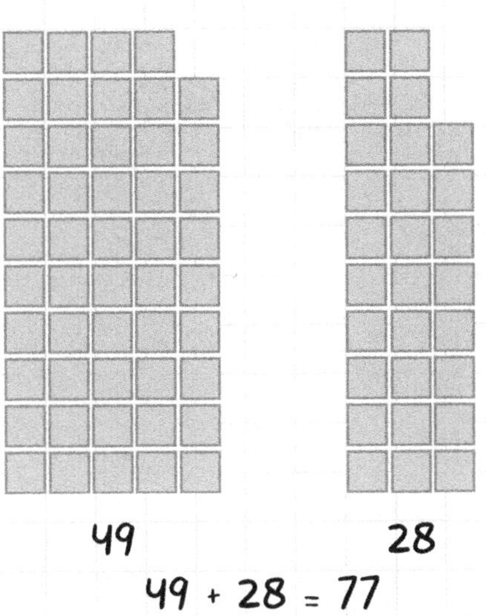

49 + 28 = 77

31. Any visual representation that depicts a strategy to show that the missing number is 62 is correct. Students can visually show 44 + 18 = 62. Students may draw representation in base 10 rods and units.

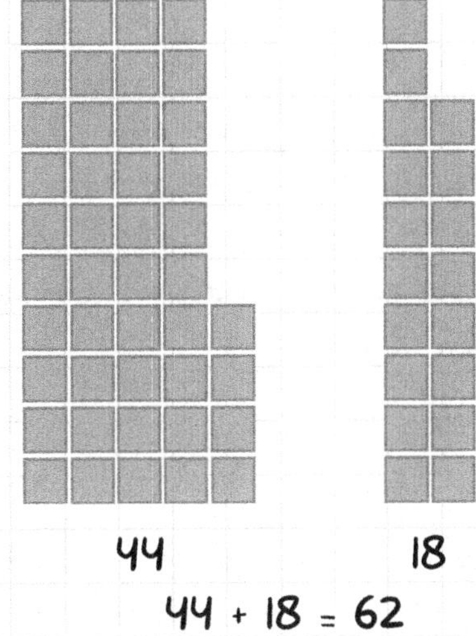

44 + 18 = 62

32. C.

The total group is 20 dogs. When 5 dogs are removed from the group, then 15 dogs remain. The subtraction equation would be (C) 20 - 5 = 15.

33. A.

The total group has 32 cookies. When 24 cookies are bought from the store, 8 cookies remain. The subtraction equation would be (A) 32 - 24 = 8.

34. Any visual representation that depicts a strategy to show that the missing number is 5 is correct. Students can visually show 25 - 5 = 20, or students can draw 25 objects and cross out 5.

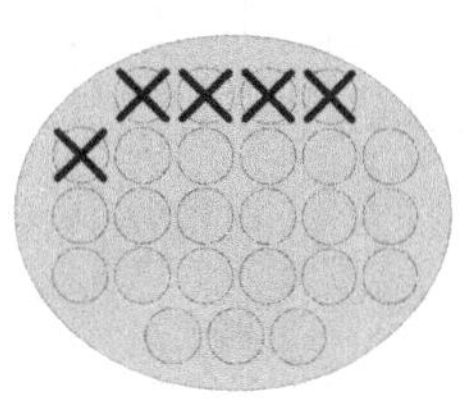

35. Any visual representation that depicts a strategy to show that the missing number is 22 is correct. Students can visually show 32 - 22 = 10, or students can draw 32 objects and cross out 10.

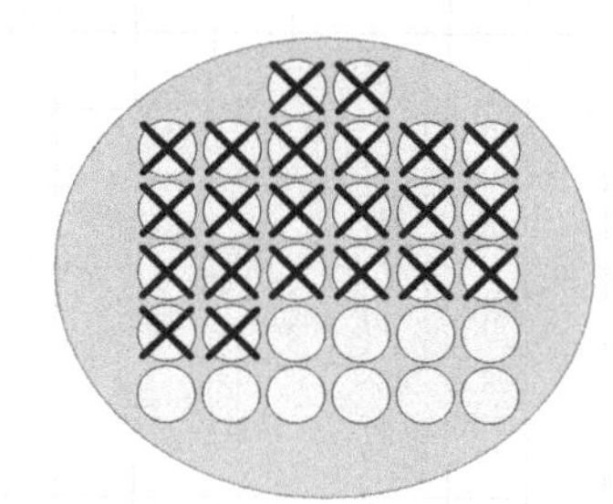

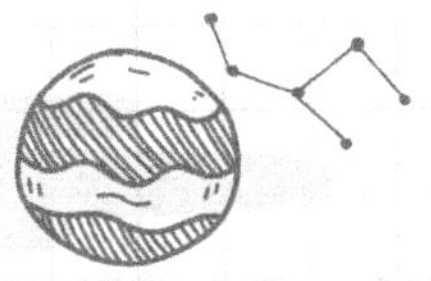

36. Any visual representation that depicts a strategy to show that the missing number is 71 is correct. Students can visually show 56 + + 15 = 71, or students can draw 15 flowers count on until they get to 71. Students may draw representation in base 10 rods and units.

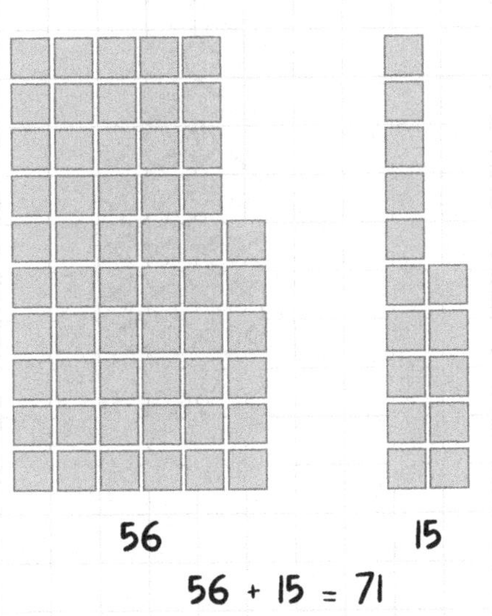

37. This is a two step problem. The first step is to see how many bugs Andrew had left after he opened the jar. He had 41 to start and then 12 flew out. The equation would be 41 - 12 = 29.

The second step is to find out how many are left after he caught 5 back. Students will add 29 + 5 = 34.

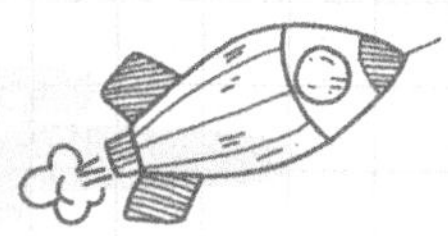

38. Any visual representation that shows that the answer is 14 is correct. Students can subtract 64 - 50 = 14, or students can draw a picture to compare and find the difference.

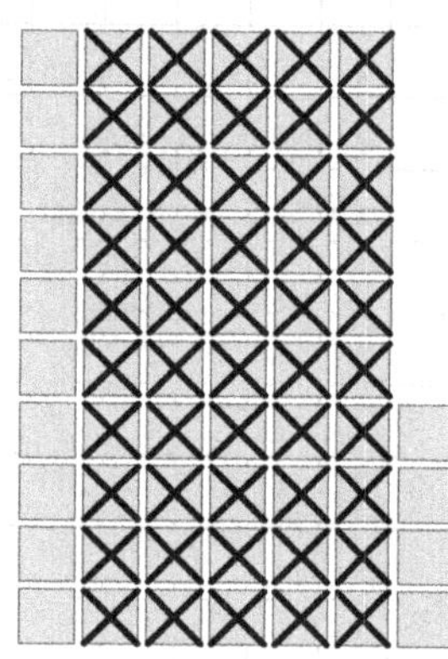

64 - 50 = 14

39. Any visual representation that shows that the answer is 9 is correct. Students can subtract 45 - 36 = 9, or students can draw a picture to compare and find the difference.

45 - 36 = 9

40. Students can subtract 65 - 12 = = 53, or students can draw a picture to compare and find the difference to show the answer is 53 lbs.

ANSWER AND EXPLANATION

SECTION 2: ADD AND SUBTRACT WITHIN 20 USING MENTAL STRATEGIES

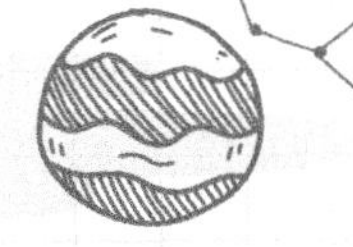

41. A.
7 + 8 = 15 is a near double addition equation. This means having a double fact that is close to this memorized, would help you solve it quickly. If you knew that 7 + 7 = 14, then you would know that adding 1 more would change the problem to 7 + 8 = 15.

42. C.
10 + 9 = 19 is a near double addition equation. This means, having a double fact that is close to this memorized, would help you solve it quickly. If you knew that 9 + 9 = 18, then you would know that adding 1 more would change the problem to 10 + 9 = 19.

43. D.
6 + 7 = 13 is a near double addition equation. This means having a double fact that is close to this memorized, would help you solve it quickly. If you knew that 7 + 7 = 14, then you would know that subtracting 1 more would change the problem to 6 + 7 = 13.

44. B.
4 + 5 = 9 is a near double addition equation. This means having a double fact that is close to this memorized, would help you solve it quickly. If you knew that 5 + 5 = 10, then you would know that subtracting 1 more would change the problem to 4 + 5 = 9.

45. Students can either add 1 to make the near double equation 6 + 7 = 13 or students can take 1 away and make near double equation 6 + 5 = 11.

46. Students can either add 1 to make the near double equation 8 + 9 = 17 or students can take 1 away and make near double equation 8 + 7 = 15.

47. Any addition equation with 2 of the same addends is correct. Examples may be 1 + 1 = 2, 2 + 2 = 4, 3 + 3 = 6, etc.

ANSWER AND EXPLANATION

SECTION 2: ADD AND SUBTRACT WITHIN 20 USING MENTAL STRATEGIES

48. Any addition equation with 2 of the same addends is correct. Examples may be 1 + 1 = 2, 2 + 2 = 4, 3 + 3 = 6, etc.

49. Any subtraction equation with 2 of the same minuend and subtrahend is correct. Examples may be 1 - 1 = 0, 2 - 2 = 0, 3 - 3 = 0, etc.

50. Any subtraction equation with 2 of the same minuend and subtrahend is correct. Examples may be 1 - 1 = 0, 2 - 2 = 0, 3 - 3 = 0, etc.

51. A.
These ten frames have 9 red and 7 yellow. 1 yellow was moved to the first ten frame to make a full 10. Now, you can visually see that 9 + 7 = 16 is the same as 10 + 6 = 16.

52. C.
These ten frames have 9 red and 4 yellow. 1 yellow was moved to the first ten frame to make a full 10. Now, you can visually see that 9 + 4 = 13 is the same as 10 + 3 = 13.

53. C.
If you take one away from the first addend (10) and add it to the second addend (7), then you change 10 + 7 to 9 + 8. Both equations have the same sum of 17.

54. D.
If you take one away from the first addend (10) and add it to the second addend (2), then you change 10 + 2 to 9 + 3. Both equations have the same sum of 12.

55. Any visual representation that shows 2 groups (10 and 6) and their equality to another 2 groups (9 and 7) is correct. Both groups should total 18. A sample picture might be:

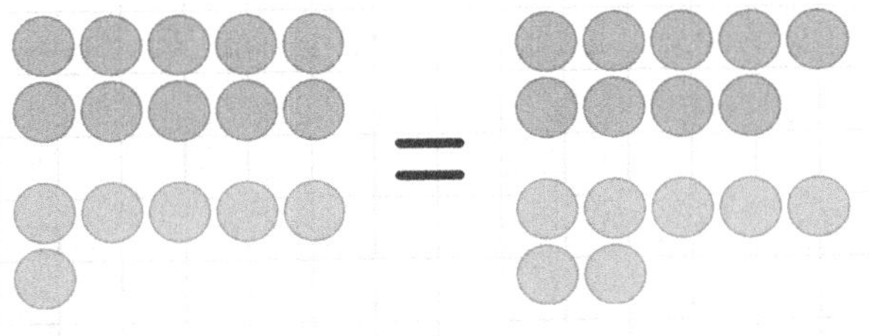

56. Any visual representation that shows 2 groups (9 and 6) and their equality to another 2 groups (10 and 5) is correct. Both groups should total 15. A sample picture might be:

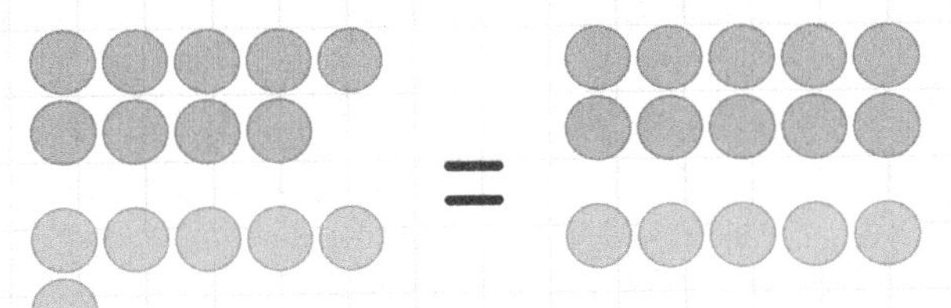

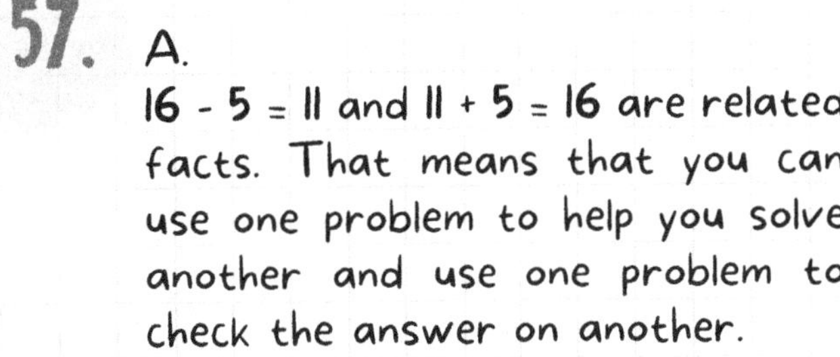

57. A.
16 - 5 = 11 and 11 + 5 = 16 are related facts. That means that you can use one problem to help you solve another and use one problem to check the answer on another.

58. C.
19 - 3 = 16 and 3 + 16 = 19 are related facts. That means that you can use one problem to help you solve another and use one problem to check the answer on another.

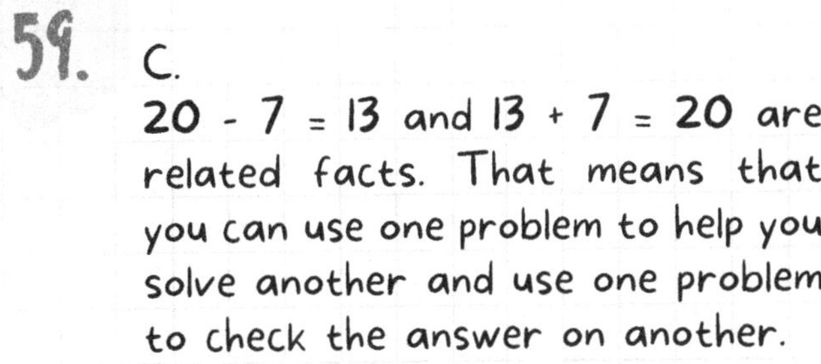

59. C.
20 - 7 = 13 and 13 + 7 = 20 are related facts. That means that you can use one problem to help you solve another and use one problem to check the answer on another.

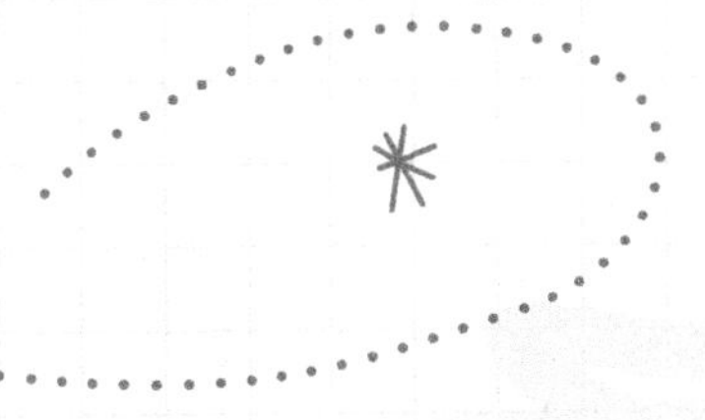

60. The student can either write 11 - 5 = 6 or 11 - 6 = 5.

61. The student can either write 19 - 2 = 17 or 19 - 17 = 2.

62. The student can either write 13 + 5 = 18 or 5 + 13 = 18.

63.

18 + 0 = 18	4 + 0 = 4	0 + 19 = 19
7 + 0 = 7	0 + 20 = 20	0 + 15 = 15

64.

12 - 0 = 12	0 - 0 = 0	17 - 0 = 17
6 - 0 = 6	10 - 0 = 10	14 - 0 = 14

65.

4 + 4 = 8	10 + 10 = 20	7 + 7 = 14
6 + 6 = 12	9 + 9 = 18	3 + 3 = 6

66.

12 - 12 = 0	2 - 2 = 0	5 - 5 = 0
1 - 1 = 0	8 - 8 = 0	7 - 7 = 0

67.

17 + 3 = 20	16 + 2 = 18	9 + 10 = 19
10 + 4 = 14	5 + 15 = 20	7 + 6 = 13
14 + 4 = 18	1 + 12 = 13	4 + 16 = 20

68.

16 - 5 = 11	17 - 2 = 15	13 - 10 = 3
12 - 4 = 8	9 - 8 = 1	14 - 6 = 8
18 - 7 = 11	19 - 12 = 7	17 - 13 = 4

ANSWER AND EXPLANATION

SECTION 2: ADD AND SUBTRACT WITHIN 20 USING MENTAL STRATEGIES

69.

8 + 12 = 20	16 + 2 = 18	20 - 10 = 10
3 + 15 = 18	8 + 8 = 16	19 - 7 = 12
17 - 7 = 10	15 - 10 = 5	2 + 13 = 15

70. B.
The missing number is 6. Students can start at 10 and count on 6 more until they reach 16. They can also do the related fact 16 - 10 = 6.

71. C.
The missing number is 9. Students can start at 9 and count on 9 more until they reach 18. They can also do the related fact 18 - 9 = 9. Knowing the double fact 9 + 9 = 18 will also help you solve this problem.

72. D.
The missing number is 7. Students can start at 8 and count on 7 more until they reach 15. They can also do the related fact 15 - 8 = 7.

73. C.
The missing number is 12. Students can start at 0 and add on 12 more until they reach 12. They can also do the related fact 12 - 0 = 12.

74. A.
The missing number is 7. Students can start at 8 and count back 7 until they reach 1. They can also do the related fact 8 - 1 = 7.

75. C.
The missing number is 6. Students can start at 12 and count back 6 until they reach 6. They can also do the related fact 12 - 6 = 6. Knowing the double fact 6 + 6 = 12 will also help you solve this.

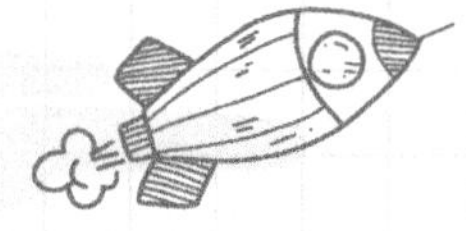

76. A.

The missing number is 18. Students can find the answer by using the related fact 10 + 8 = 18. Also, knowing the addition facts with 10 will help solve this problem.

77. D.

The missing number is 13. Students can find the answer by using the related fact 7 + 6 = 13. Also, knowing that 6 + 6 = 12 will help because students can figure out near double fact is 7 + 6.

78.

15 + 5 = 20	5 + 13 = 18	5 + 6 = 11
2 + 8 = 10	7 + 7 = 14	17 + 0 = 17

79. To find how many muffins Donna has left, students will start with 12 and subtract 6: 12 - 6 = 6. Donna has 6 muffins. To find out how many muffins Matt has, students start with 6 and subtract the 1 he ate: 6 - 1 = 5. Donna has 6 and Matt has 5.

80. To find how many apples I have left, I need to start with 14 and subtract 10: 14 - 10 = 4. Then, I need to take the difference, (4), and subtract 1: 4 - 1 = 3. There are 3 apples left.

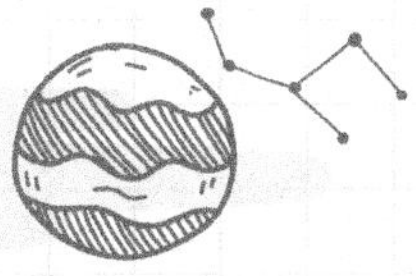

81. There are 13 apples. 13 is an odd number because 1 apple is without a partner.

82. There are 18 watermelon slices. 18 is an even number because all the slices have a partner.

83. There are 24 bananas. 24 is an even number because all the bananas have a partner.

84. There are 17 oranges. 17 is an odd number because 1 orange is without a partner.

85. Students can pick any single digit odd number (1, 3, 5, 7, 9). A sample illustration may be:

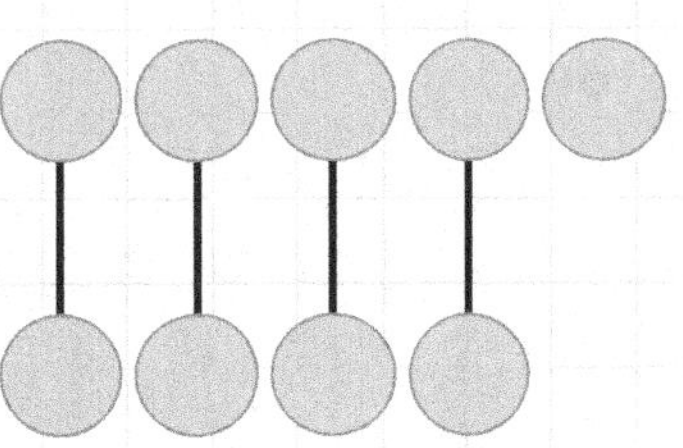

86. Students can pick any single digit even number (2, 4, 6, 8). A sample illustration may be:

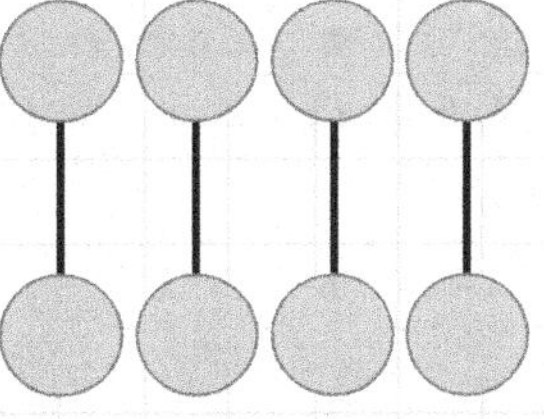

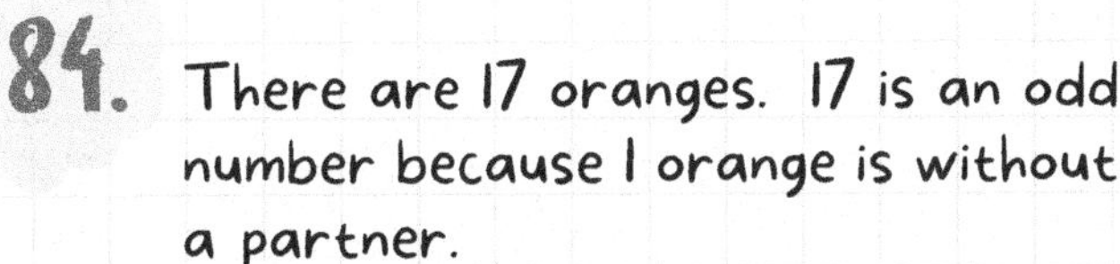

87. Students can pick any double digit odd number. An odd double digit number is a number that ends in 1, 3, 5, 7, 9. A sample illustration may be:

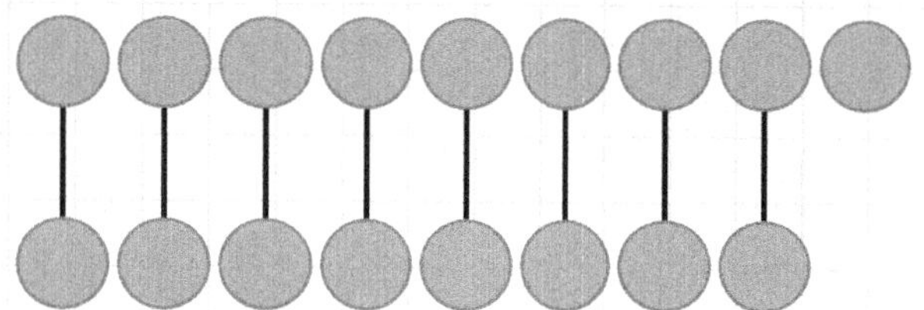

88. Students can pick any double digit even number. An even double digit number is a number that ends in 2, 4, 6, 8, 0. A sample illustration may be:

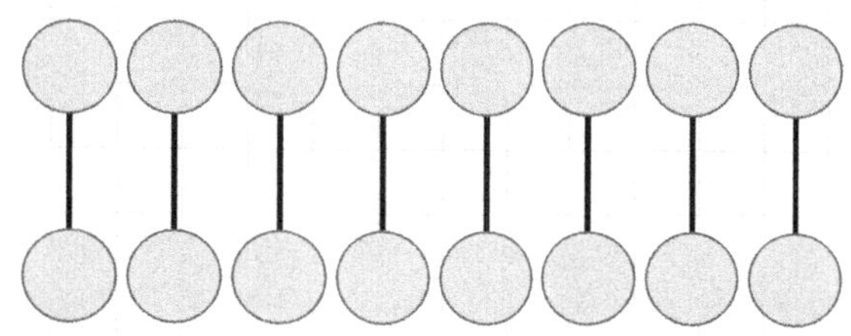

89. 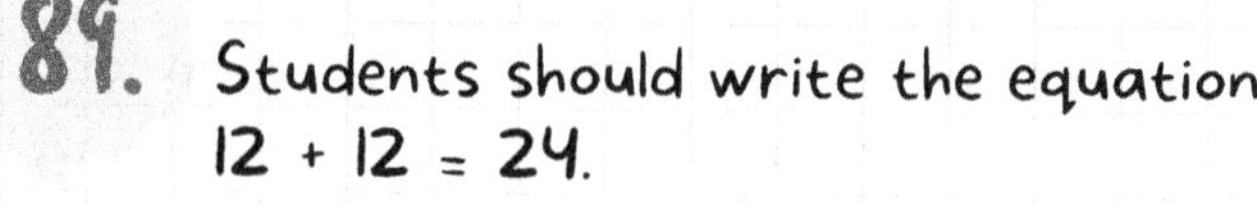
Students should write the equation 12 + 12 = 24.

90. Students should write the equation 8 + 8 = 16.

91. B.
The picture for (B) has 4 circles on the top and 4 circles on the bottom. This shows the equation 4 + 4 = 8.

92. C
The picture for (C) has 7 circles on the top and 7 circles on the bottom. This shows the equation 7 + 7 = 14.

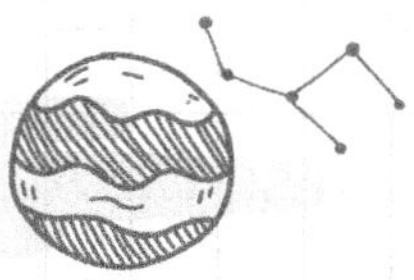

93. Any visual representation that shows 10 eyes is correct. For example:

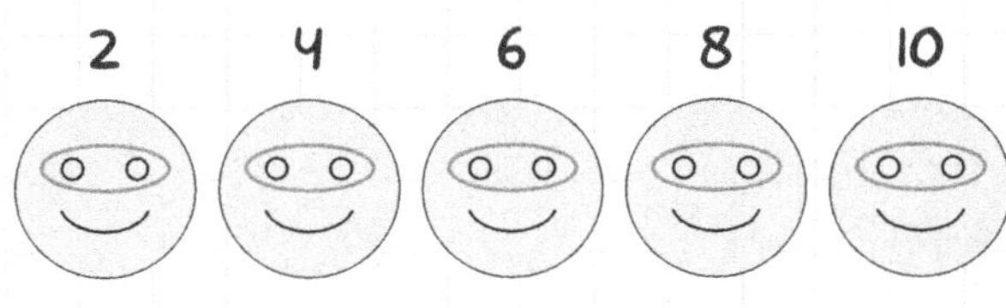

94. Any visual representation that shows 24 feet is correct. For example:

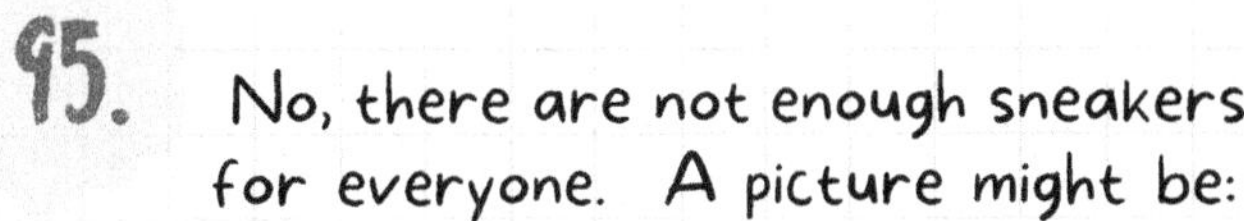

95. No, there are not enough sneakers for everyone. A picture might be:

96. There are enough mittens for everyone. A picture might be:

97.

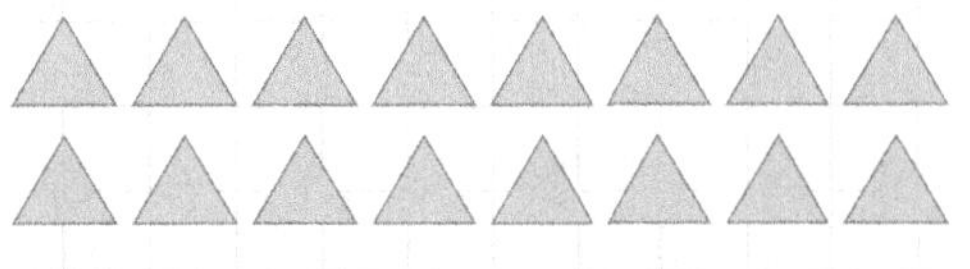

98.

99. 10

100. 15

101. 3 + 3 + 3 = 9

102. 5 + 5 + 5 + 5 = 20

103. 3 + 3 + 3 + 3 + 3 = 15

104. 4 + 4 + 4 = 12

105. 4 + 4 + 4 + 4 = 16

106. 2 + 2 + 2 + 2 + 2 = 10

107. 3 + 3 + 3 + 3 = 12

108. 4 + 4 + 4 + 4 + 4 = 20

109. 2 + 2 + 2 = 6

110. 5 + 5 + 5 = 15

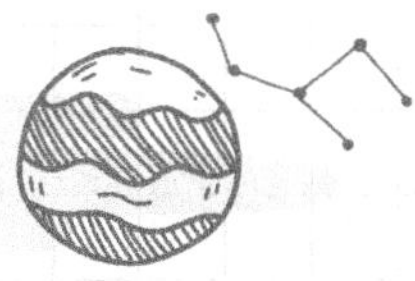

111. Any visual representation that shows 4 objects in 3 rows is correct. For example:

112. Any visual representation that shows 2 objects in 4 rows is correct. For example:

113. Any visual representation that shows 3 objects in 5 rows is correct. For example:

114. Any visual representation that shows 5 objects in 3 rows is correct. For example:

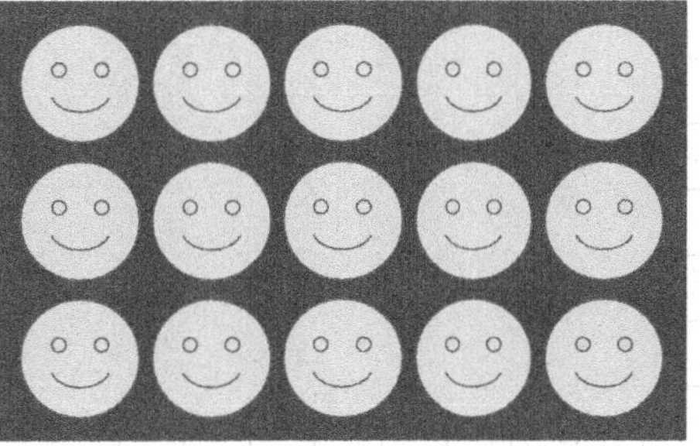

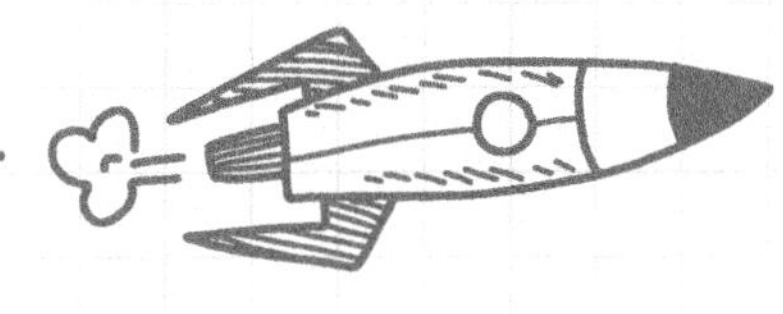

ANSWER AND EXPLANATION

SECTION 3: FOUNDATIONS FOR MULTIPLICATION

115. Any visual representation. It must be done with 4 objects in 4 rows. For example:

116. 8 + 8 + 8 = 24

117. 5 + 5 + 5 + 5 = 20

118. Students can wither write 2 + 2 + 2 + 2 + 2 + 2 + 2 + 2 = 16 or 4 + 4 + 4 + 4 = 16

119. 2 + 2 + 2 + 2 + 2 + 2 + 2 = 14

120. Students can write 3 + 3 + 3 + 3 = 12 or 6 + 6 = 12

ANSWER AND EXPLANATION

SECTION 1: PLACE VALUE

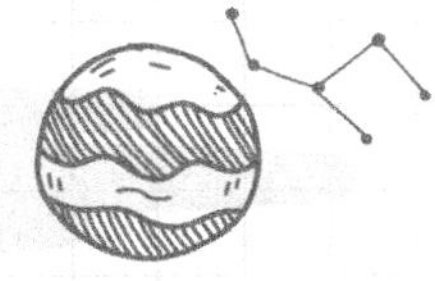

1. There are 7 hundreds in the number 706.

2. There are 5 tens in the number 859.

3. There are 3 tens in the number 632.

4. There are 0 ones in the number 500.

5. There are 4 hundreds in the number 498.

6. There are 7 ones in the number 257.

7. A.
There are 4 rods of 10 blocks each. If you count by 10s, you get 40 in all.

8. B.
There are 8 rods of 10 blocks each. If you count by 10s, you get 80 in all.

9. C.
There are 10 rods of 10 blocks each. If you count by 10s, you get 100 in all.

10. D.
The number 700 has 7, hundreds 0 tens, and 0 ones.

ANSWER AND EXPLANATION

SECTION 1: PLACE VALUE

11. B.
The number 843 has 8 hundreds, 4 tens, and 3 ones.

12. B.
The number 601 has 6 hundreds, 0 tens, and 1 ones.

13. 300

14. 400

15. 900

16. 700

17. 35, 40, 45, 50, 55, 60, 65, 70

18. 140, 150, 160, 170, 180, 190, 200, 210

19. 560, 565, 570, 575, 580, 585, 590, 595

20. 200, 300, 400, 500, 600, 700, 800, 900

21. 850, 855, 860, 865, 870, 875, 880, 885

22. 710, 720, 730, 740, 750, 760, 770, 780

23. 140, 240, 340, 440, 540, 640, 740, 840

ANSWER AND EXPLANATION

SECTION 1: PLACE VALUE

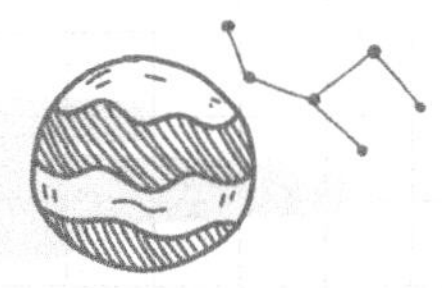

24. A.
The number 679 has 6 hundreds, 7 tens, and 9 ones. When written in expanded form it is: 679 = 600 + + 70 + 9.

25. C.
The number 905 has 9 hundreds, 0 tens, and 5 ones. When written in expanded form it is: 905 = 900 + 0 + 5.

26. D.
The number 260 has 2 hundreds, 6 tens, and 0 ones. When written in expanded form it is: 260 = 200 + 60 + 0.

27. The number 842 has 8 hundreds, 4 tens, and 2 ones. When written in expanded form it is: 842 = 800 + 40 + 2.

28. The number 502 has 5 hundreds, 0 tens, and 2 ones. When written in expanded form it is: 502 = 500 + 0 + 2.

29. The number 198 has 1 hundred, 9 tens, and 8 ones. When written in expanded form it is: 198 = 100 + 90 + 8 ones.

30. 520 < 945 If you compare the hundreds, 520 has 5 hundreds, and 945 has 9 hundreds. 5 hundreds is less than 9 hundreds.

31. 490 < 590 If you compare the hundreds, 490 has 4 hundreds, and 590 has 5 hundreds. 4 hundreds is less than 5 hundreds.

ANSWER AND EXPLANATION

SECTION 1: PLACE VALUE

32. 398 > 397 These two numbers have equal amounts of hundred and tens. If you compare the ones, 8 is greater than 7.

33. 621 = 621 These numbers have equal hundreds, tens, and ones, so they are equal to each other.

34. 764 > 729 These numbers have equal hundred. If you compare the tens, 764 has 6 tens, and 729 has 2 tens. 6 tens is greater than 2 tens.

35. 542 = 542 These numbers have equal hundreds, tens, and ones, so they are equal to each other.

36. Students can write any number that is greater than 643. Some examples might be 644, 659, 720.

37. Students should write the number 982. These numbers have the same amount of hundreds, tens, and ones.

38. Students can write any number that is less than 201. Some examples might be 200, 197, 8.

39. Students should write the number 399. These numbers have the same amount of hundreds, tens, and ones.

40. Students can write any number that is less than 610. Some examples might be 609, 600, 432.

ANSWER AND EXPLANATION

SECTION 2: USING PLACE VALUE AND PROPERTIES OF OPERATIONS TO ADD AND SUBTRACT

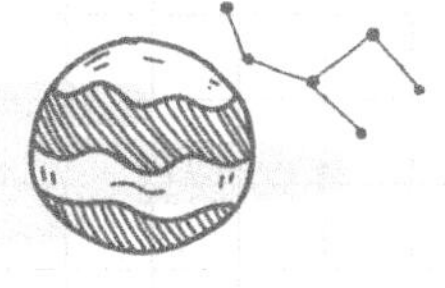

41.

68 + 32 = 100	21 + 65 = 86	33 + 60 = 93
19 + 40 = 59	52 + 38 = 90	77 + 21 = 98

42.

53 - 27 = 26	96 - 43 = 53	84 - 31 = 53
25 - 19 = 6	34 - 33 = 1	63 - 50 = 13

43. Students should draw:

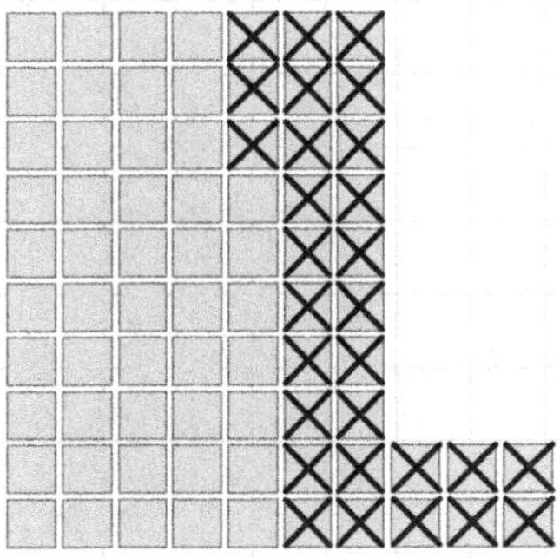

The difference should be 47.

44. Students should draw:

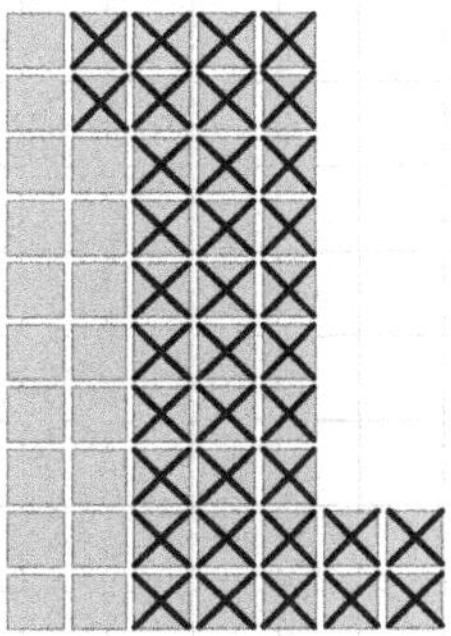

The difference should be 18.

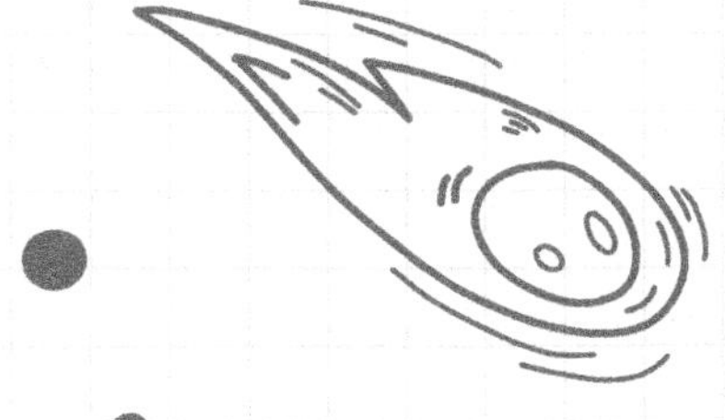

ANSWER AND EXPLANATION

SECTION 2: USING PLACE VALUE AND PROPERTIES OF OPERATIONS TO ADD AND SUBTRACT

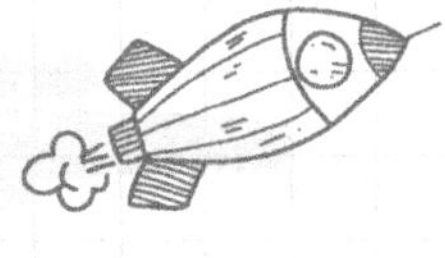

45. Students should draw:

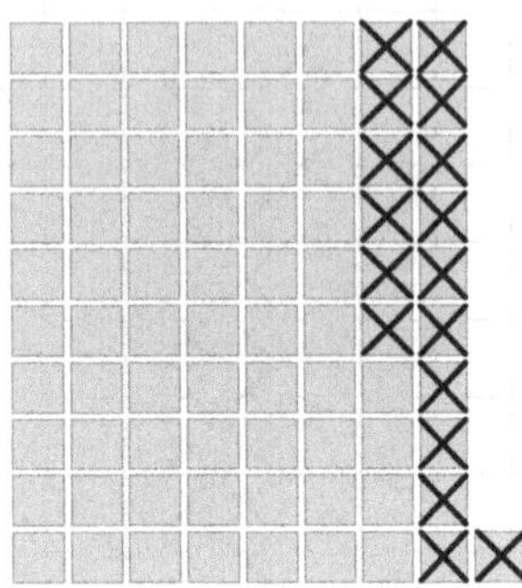

The difference should be 64.

46. Students should draw:

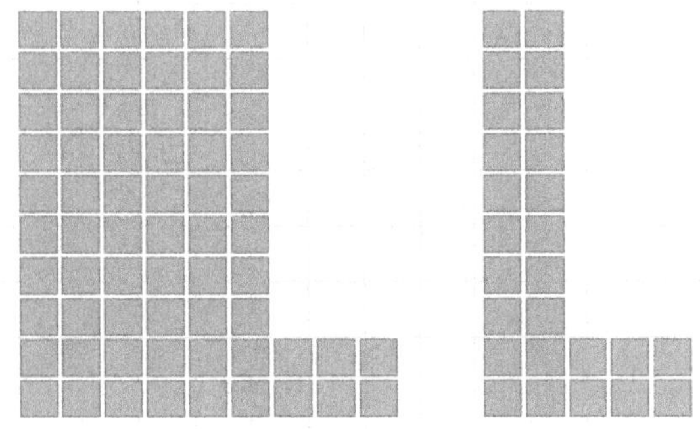

The sum should be 92.

47. Students should draw:

The sum should be 97.

48. Students should find the sum:
30 + 15 + 11 + 25 = 81
To show their work, students can draw a picture using ten rods and units. They can show how they added the tens and ones together or students can visually depict their addition strategy.
An example may be:
Adding the ones 0 + 5 + 1 + 5 = 11
Adding the tens 30 + 10 + 10 + 20 = = 70
Adding the tens and ones: 70 + 1 = 81

49. 89

To show their work, students can draw a picture using ten rods and units. They can show how they added they tens and ones together, or students can visually depict their addition strategy.

An example may be:

Adding the ones 1 + 6 + 2 + 0 = 9

Adding the tens 50 + 0 + 10 + 20 = = 80

Adding the tens and ones: 80 + 9 = 89

50. 99

To show their work, students can draw a picture using ten rods and units. They can show how they added the tens and ones together or students can visually depict their addition strategy.

An example may be:

Adding the ones 6 + 2 + 2 + 9 = 19

Adding the tens 20 + 50 + 0 + 10 = = 80

Adding the tens and ones: 80 + 19 = 99

51. 81

To show their work, students can draw a picture using ten rods and units. They can show how they added the tens and ones together or students can visually depict their addition strategy.

An example may be:

Adding the ones 8 + 7 + 1 + 5 = 21

Adding the tens 30 + 10 + 20 = = 60

Adding the tens and ones: 60 + 21 = 81

52. C.

First add the ones (2 + 1 = 3). Then, add the tens (4 + 5 = 9). Last, add the hundreds (5 + 3 = 8). The sum is 893.

53. A.

First add the ones (3 + 0 = 3). Then, add the tens (6 + 1 = 7). Last, add the hundreds (2 + 7 = 9). The sum is 973.

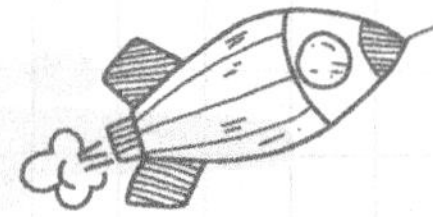

54. C.

First add the ones (5 + 1 = 6). Then, add the tens (8 + 5 = 13). Since the tens have a sum that is greater than 10, you need to add an extra 10 to the hundreds. Then, add the hundreds (4 + 2 + 1 = 7). The sum is 736.

55. D.

First add the ones (9 + 5 = 14). Since the ones have a sum that is greater than 10, you need to add ten to the tens. Then, add the tens (6 + 4 + 1 = 11). Since the tens have a sum that is greater than 10, you need to add 1 extra to the hundreds. Then, add the hundreds (5 + 3 + 1 = 9). The sum is 914.

56.

642 + 261 = 903	111 + 601 = 712	300 + 456 = 903
705 + 194 = 899	471 + 222 = 683	644 + 110 = 754

57.

597 + 213 = 810	327 + 403 = 730	685 + 126 = 811
476 + 415 = 891	618 + 105 = 723	700 + 204 = 904

58. D.

First subtract the ones (7 - 2 = 5). Then, subtract the tens (6 - 1 = 5). Last, subtract the hundreds (3 - 1 = = 2). The difference is is 215.

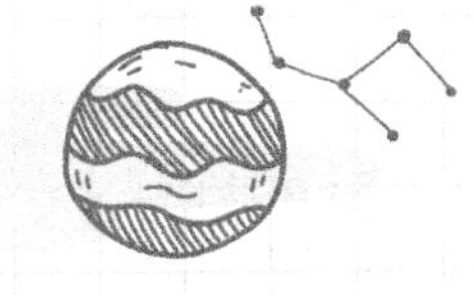

59. A.
First subtract the ones (5 - 0 = 5). Then, subtract the tens (3 - 1 = 2). Last, subtract the hundreds (9 - 7 = = 2). The difference is is 225.

60. C.
First, subtract the ones. Since there are not enough ones to subtract from, you need to borrow 10 from the tens (7 + 10 = 17), then subtract the ones (17 - 9 = 8). Since you took a ten away from the tens column, it is now 8 - 5 = = 3. Then, subtract the hundreds (3 - 2 = 1). The difference is 138.

61. D.
First subtract the ones. Since there are not enough ones to subtract from, you need to borrow 10 from the tens (5 + 10 = 15), then subtract the ones (15 - 6 = 9). Since you took a ten away from the tens column it is now 7 - 6 = = 1. Then, subtract the hundreds (9 - 5 = 4). The difference is 419.

62.

721 - 320 = 401	544 - 131 = 413	674 - 265 = 409
367 - 205 = 162	493 - 380 = 113	823 - 603 = 220

63.

580 - 190 = 390	642 - 225 = 417	210 - 155 = 55
735 - 457 = 278	427 - 159 = 268	351 - 207 = 144

64. Students should write the addition equation 47 + 10 = 57 or 10 + 47 = 57.

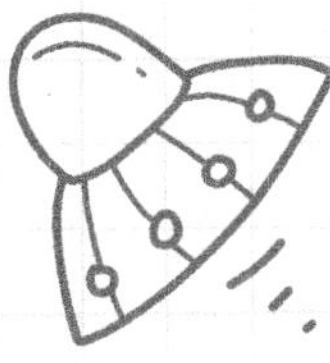

65. Students should write the addition equation 645 + 100 = 745 or 100 + 645 = 745.

66. Students should write the subtraction equation 278 - 10 = = 268.

67. Students should write the subtraction equation 851 - 100 = = 751.

68. Students should write the addition equation 390 + 10 = 400 or 10 + 390 = 400.

69. Students should write the subtraction equation 529 - 100 = = 429.

70. 199 cookies

Because this problem is asking you to combine all of her cookies to form a total, you need to write an addition equation. The addition equation should be 34 + 25 + 40 = = 99. Students should add the ones (4 + 5 = 9) and add the tens (3 + 2 + 4 = 9). Since the bake sale added 100 more cookies, students need to add 100 more to the total: 99 + 100 = 199 cookies in all.

71. $77

To solve this problem, students need to write a subtraction equation. Students should start with the price of the bike and subtract the amount Emma has in order to see what's left. The subtraction equation should be: $200 - $123 = $77. Emma still needs $77 to buy the bike.

ANSWER AND EXPLANATION

SECTION 2: USING PLACE VALUE AND PROPERTIES OF OPERATIONS TO ADD AND SUBTRACT

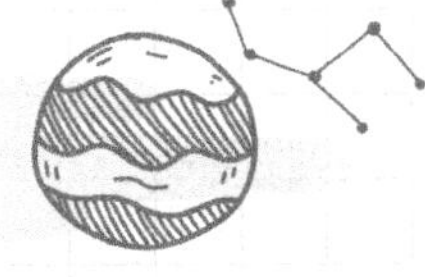

72. 90 pennies
To solve the problem, students need to subtract 10 from 100. The subtraction equation would be 100 - 10 = 90 pennies. Parker has 90 pennies.

73. 126 apples
To solve the problem, students need to subtract 24 from 150. The subtraction equation would be 150 - 24 = 126 apples. There are 126 apples left in the barrel.

74. 656 pom poms
Because this problem is asking you to combine all of the pom poms to form a total, you need to write an addition equation. The addition equation should be 367 + 289 = 656 pom poms. The art teacher needs 656 pom poms in all.

75. 32 years older
Students need to use subtraction to compare the ages of the dad and grandpa. Students should write: 72 - 40 = 32. His grandpa is 32 years older than his dad.

76. 68 pounds
To solve this problem, students need to add 10 pounds to the original weight of the dog. The addition equation should be 58 + 10 = 68 pounds. The dog weighs 68 pounds now.

77. This is a two step problem. In the first step, the students need to compare 84 and 97. If you compare the tens, then 9 tens is more than 8 tens, so 97 is greater which means Dee has more dolls. For the second step, students need to use subtraction to compare the amount of dolls to find out how many more Dee has than Jess. The subtraction equation should be 97 - 84 = 13 dolls. Dee has 13 more dolls than Jess.

ANSWER AND EXPLANATION

SECTION 2: USING PLACE VALUE AND PROPERTIES OF OPERATIONS TO ADD AND SUBTRACT

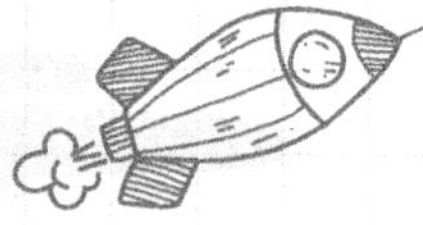

78. 89 stamps

Because this problem is asking you to combine all of her stamps to form a total, you need to write an addition equation. The addition equation should be 19 + 37 + 28 + 5 = 89. Students should add the ones (9 + 7 + 8 + 5 = = 29) and add the tens (1 + 3 + 2 = 6). When you combine the tens and ones, the addition equation would be 60 + 29 = 89 stamps. Natalie has 89 stamps in all.

79. 7 crayons missing

To solve the problem, students need to write a subtraction equation. They will start with the total number of crayons (64) and then subtract the crayons they found (57). The subtraction equation should be: 64 - 57 = 7 crayons. They are still missing 7 crayons.

80. There are enough tickets left.

To solve the problem, students need to write a subtraction equation. Start with the total number of seats (735) and subtract the seats that were sold (729) to find out how many seats are still available. The subtraction equation should be: 735 - 729 = 6 seats. David's family has 5 people. Since 5 is less than 6, there are enough tickets for them to go to the game.

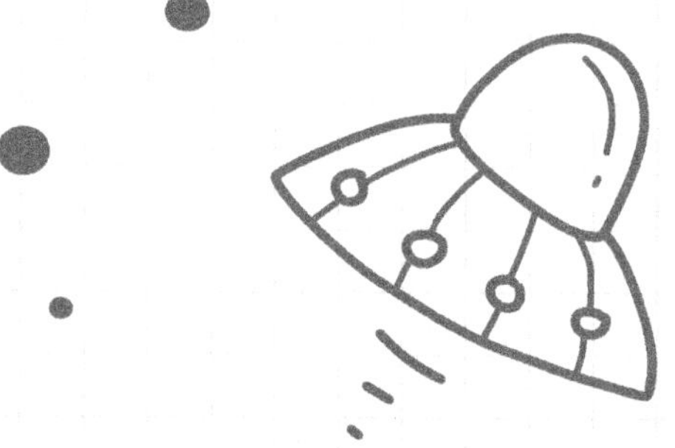

ANSWER AND EXPLANATION

SECTION 1: MEASURING IN STANDARD UNITS

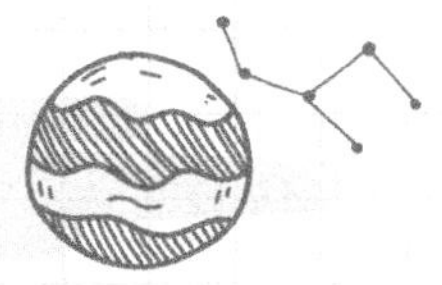

1. B.
A measuring tape would be the most efficient tool because it can measure in feet and it can extend the length of the couch.

2. C.
A ruler would be the most efficient tool because a picture frame is a small object and a ruler can measure it in inches.

3. C.
A yardstick would be the most efficient tool because it is long enough to measure the length of the toybox and a yard is a standard unit that is close to the length of a toybox.

4. B.
You would measure the length of a book in inches because a book is a small object and an inch is a small unit of measurement.

5. D.
You would measure the height of a door in feet because a door is a tall object and a foot is a unit of measurement that corresponds to the height of a door.

6. From nose to tail, the shark is approximately 6 inches long.

7. From head to end, the caterpillar is approximately 5 centimeters long.

8. From nose to tail, the fish is approximately 10 centimeters long.

9. From end to end, the pillow is approximately 3 inches long.

10. From heel to toe, the sneaker is approximately 13 centimeters long.

11. To solve the problem, students need to write a subtraction equation to compare the inches to paperclips. The subtraction equation should be **48 - 36 = 12** inches. It takes **12** more inches than paper clips to measure the table.

12. To solve the problem, students need to write a subtraction equation to compare the inches to paperclips. The subtraction equation should be **65 - 50 = 15** paper clips. It takes **15** fewer paperclips than inches to measure the television.

13. To solve the problem, students need to write a subtraction equation to compare the centimeters to toothpicks. The subtraction equation should be **106 - 18 = 88** centimeters. It takes **88** more centimeters than toothpicks to measure the bat.

14. To solve the problem, students need to write a subtraction equation to compare the centimeters to toothpicks. The subtraction equation should be **152 - 26 = 126** toothpicks. It takes **126** fewer toothpicks than centimeters to measure the rug.

15. To solve the problem, students need to write a subtraction equation to compare the feet to tiles. The subtraction equation should be **6 - 4 = 2** feet. It takes **2** fewer feet than tiles to measure her height.

16. To solve the problem, students need to write a subtraction equation to compare the feet to tiles. The subtraction equation should be **8 - 5 = 3** tiles. It takes **3** more tiles than feet to measure Lindsey's height.

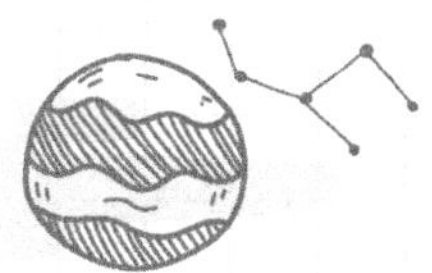

17. A.

Since it took fewer paperclips to measure the book than inches, that means that the paperclips are bigger than an inch, so it takes more inches than paperclips to measure the book.

18. C.

Since it took fewer inches to measure Pam's height than centimeters, that means centimeters are smaller than inches, so it takes more centimeters than inches to measure Pam's height.

19. B.

Since it took fewer feet to measure Jim's bed than inches, that means inches are smaller than feet, so it takes fewer feet than inches to measure Jim's bed.

20. D.

Since it took fewer yards to measure Kelly's bedroom than feet, that means yards are bigger than feet, so it takes fewer yards than feet to measure Kelly's bedroom.

21. B.

The shorter fence is approximately half the length of the longer fence. Since the longer fence is 10 meters long, then the shorter fence is approximately 5 meters because 5 is half of 10.

22. C.

The longer bed is only a little bit longer than the shorter bed. If the shorter bed is 4 feet, then the answer choice that is a little bigger than 4 feet is 6 feet.

ANSWER AND EXPLANATION

SECTION 1: MEASURING IN STANDARD UNITS

23. B.
The shorter door is approximately half the height of the taller door. Since the shorter door is 2 meters tall, the taller door is approximately 4 meters because 2 is half of 4.

24. D.
The longer bench is only a little bit longer than the shorter bench. If the shorter bench is 48 inches, then the answer choice that is a little bigger than 48 inches is 60 inches.

25. C.
The longer eraser is only a little bit longer than the shorter eraser. If the longer eraser is 8 centimeters, then the answer choice that is a little shorter than 8 centimeters is 5 centimeters.

26. A.
6 inches is the only answer choice that makes sense in regards to the size of a cell phone. The rest of the options would be too small or too big.

27. C
4 centimeters is the only answer choice that makes sense in regards to the size of a crayon. The rest of the options would be too small or too big.

28. A
10 feet is the only answer choice that makes sense in regards to the height of a basketball hoop. The rest of the options would be too small or too big.

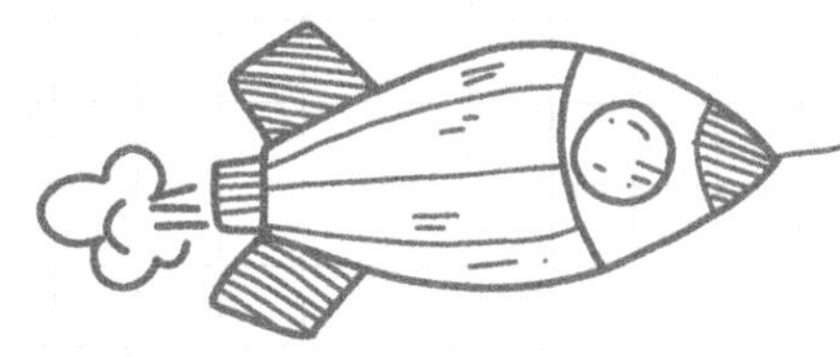

ANSWER AND EXPLANATION

SECTION 1: MEASURING IN STANDARD UNITS

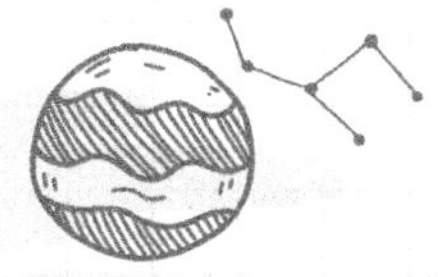

29. C.
7 centimeters is the only answer choice that makes sense in regards to the width of a dollar bill. The rest of the options would be too small or too big.

30. D.
25 meters is the only answer choice that makes sense in regards to the length of a pool. The rest of the options would be too small or too big.

31. To solve the problem, students need to write a subtraction equation to compare the lengths of the jump ropes. The subtraction equation would be: 5 - 4 = 1 foot. Rose's jump rope is 1 foot longer than Avery's jump rope.

32. To solve the problem, students need to write a subtraction equation to compare the lengths of the races. The subtraction equation would be: 50 - 25 = 25 yards. Justin is running 25 yards more than Jimmy.

33. To solve the problem, students need to write a subtraction equation to compare the lengths of the boards. The subtraction equation would be: 12 - 9 = 3 feet. The brother's board was 3 feet shorter.

34. To solve the problem, students need to write a subtraction equation to compare the carpet length to the length of the room. The subtraction equation would be: 13 - 6 = 7 feet. Jackie will have 7 feet of carpet left over.

35. Nate has to use the basement door because 72 is greater than 60. To figure out how much extra space Nate will have, you need to write a subtraction sentence to compare the width of the couch to the width of the door. The subtraction equation would be 72 - 60 = 12 inches. Nate will have 12 inches of extra room to get through the door.

36. To solve the problem, students need to write a subtraction equation to compare the lengths of the leashes. The subtraction equation would be: 182 - 119 = 63cm. Her new leash is 63cm longer than the old leash.

37. To solve the problem, students need to write a subtraction equation to compare the lengths of the yarn. The subtraction equation would be: 30 - 24 = 6 inches. She would need to trim off 6 inches in order to give her teacher a piece of yarn that was 24 inches long.

38. To solve the problem, students need to write a subtraction equation to compare the lengths of the pencils. The subtraction equation would be: 18 - 4 = 14 cm. Jill sharpened away 14cm of her pencil between Monday and Friday.

39. To solve the problem, students need to write a subtraction equation to compare the lengths of the table and tablecloth. The subtraction equation would be: 100 - 64 = 36 inches. The tablecloth is 36 inches longer than the table.

40. To solve the problem, students need to write a subtraction equation to compare the height of the sunflowers. The subtraction equation would be: 84 - 76 = 8 inches. The mom's sunflower is 8 inches taller than the daughter's sunflower.

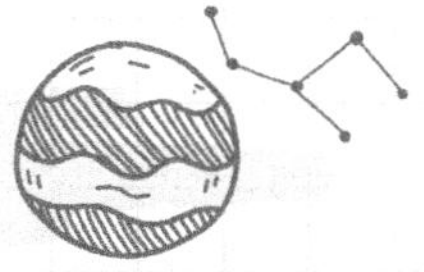

41. To solve this problem, students should write an addition equation. The hose stretches to her garden, which is 38 feet and then an additional 10 feet. To solve, students should write the addition equation: 38 + 10 = 48 feet.

42. To solve the problem, students need to write a subtraction equation to compare distances walked. The subtraction equation would be: 10 - 6 = 4 feet. The neighbor walked 4 more feet.

43. This is a problem with an unknown number. To solve, students can either subtract 15 - 10 = 5 or students can start at 10 and count on 5 more to get to 15. The answer should be that Mrs. MacDonald's rope is 5 feet long.

44. A sample representation may be:

10 feet	10 feet	10 feet	10 feet

10 + 10 + 10 + 10 = 40 feet

45. To solve the problem, students need to write a subtraction equation to compare height between a baby giraffe and an adult giraffe. The subtraction equation would be: 18 - 6 = 12 feet. Giraffes grow 12 feet from being a baby to being an adult.

46. To solve this problem, students need to combine the 2 parts of the fence; the painted and unpainted parts. Students should write the addition equation: 25 + 14 = 39 feet. Hannah fence is 39 feet long.

ANSWER AND EXPLANATION

SECTION 2: ADDITION AND SUBTRACTION USING LENGTH

47. This problem has many possible solutions. Students can write any addition equation in which the 2 addends (one for the pink yarn and one for the yellow yarn) equals 210. Some example equations may be: 200 + 10 = 210, 120 + 90 = 210, 100 + 110 = 210.

48. To solve the problem, students need to write a subtraction equation to compare lengths of the jumps of the two different animals. The subtraction equation would be: 25 - 10 = 15 feet. A kangaroo can jump 15 feet more than a hare.

49. Students can solve this problem by taking the total length of the line (32 feet) and subtracting the girls' length (18 feet) to determine the length of the boys' line. The subtraction equation would be 32 - 18 = 14 feet. Students can also start at 18 feet and count on 14 more until they get to 32. The answer should be: The boys' line is 14 feet long.

50. Students can solve this problem by taking the total length of the hot dog (12 inches) and subtracting the length that was remaining (5 inches) to determine the length that was eaten. The subtraction equation would be 12 - 5 = 7 inches. Students can also start 5 inches and count on 7 more until they get to 12. The answer should be: 7 inches of the hot dog were eaten.

51. Students need to add the lengths of the fence and write an addition equation: 7 + 7 + 7 + ? = 28.
To solve, students can add the known addends and get 21 and then subtract 28 - 21 = 7 feet to figure out the unknown addend. Frank needs 7 more feet of fence.

ANSWER AND EXPLANATION

SECTION 2: ADDITION AND SUBTRACTION USING LENGTH

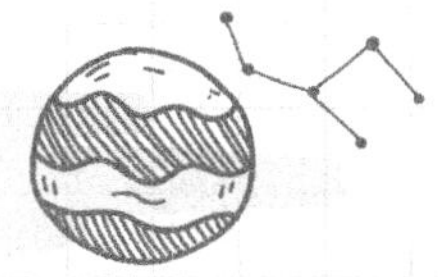

52. Students should draw a line that looks like this:

4 inches | 2 inches

4 + 2 = 6 inches

53. Students should draw a line that looks like this:

1 inch | 4 inches

1 + 4 = 5 inches

54. Students should draw a line that looks like this:

3 inches | 2 inches

3 + 2 = 5 inches

55. Students should cut the line as shown and write the corresponding subtraction equation:

6 - 2 = 4 inches

56. Students should cut the line as shown and write the corresponding subtraction equation:

5 - 3 = 2 inches

57. Students should draw a picture that looks like this:

13 feet

13 feet

13 feet

13 + 13 + 13 = 39 feet

58. Students should draw a picture that looks like this:

14 inches

14 inches

14 inches

14 inches

14 + 14 + 14 + 14 = 56 inches

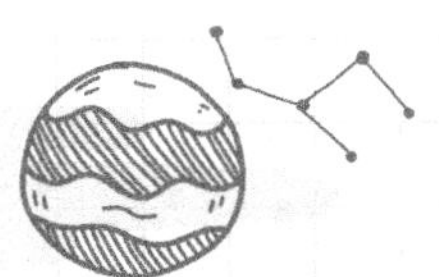

59. Students should draw a picture that looks like this:

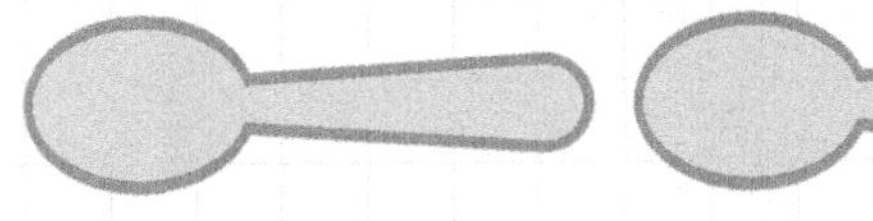

18 centimeters 18 centimeters

18 + 18 = 36 centimeters

60. Students should draw a picture that looks like this:

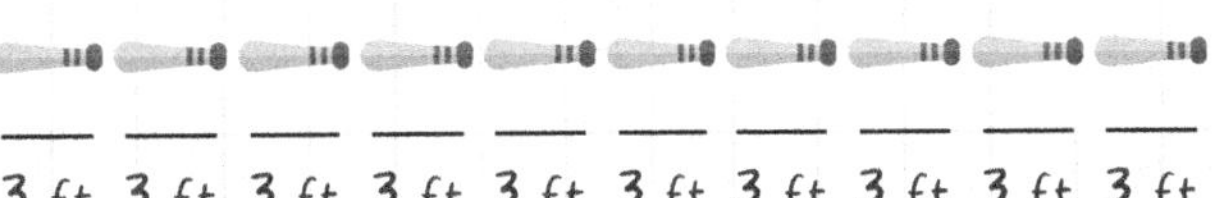

Students may mentally know that if you have 10 items 3 ft long then the total is 30 feet. Or they can write and addition equation and solve: 3 + 3 + 3 + 3 + 3 + 3 + 3 + 3 + + 3 + 3 = 30 feet.

61.

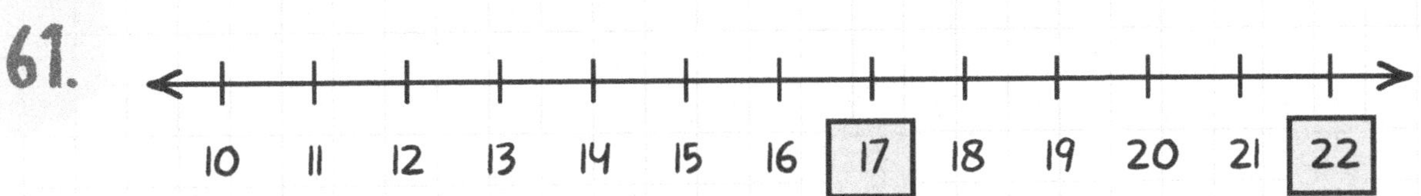

62.

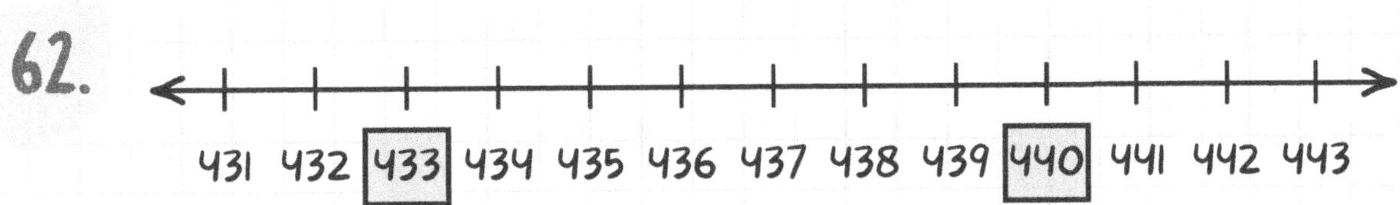

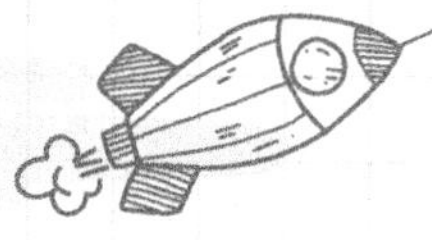

63.

691 692 693 694 [695] 696 697 698 699 [700] [701] 702 703

64.

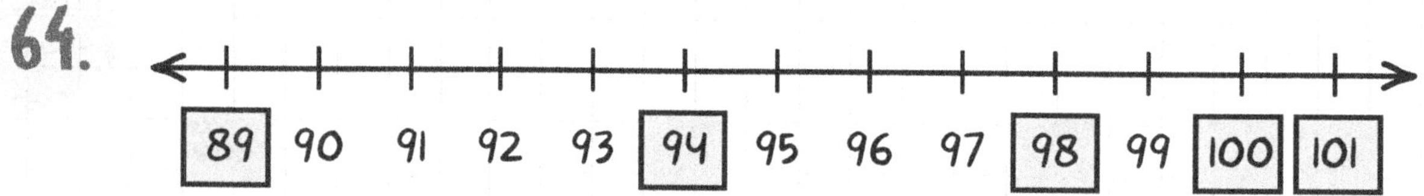

65. The number 25 should be placed closer to the 27 than to the 20.

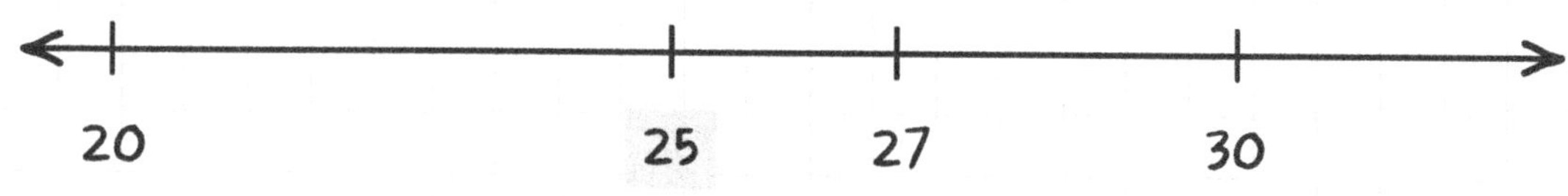

66. The number 301 should be placed right after 300.

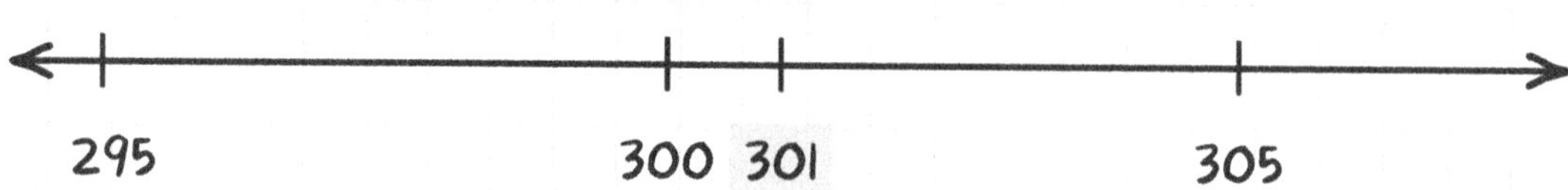

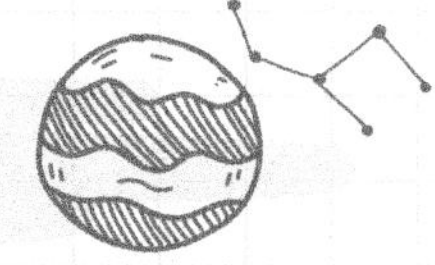

67. The number **65** should be close to the beginning of the number line.

68. The number **867** should be closer to **870** than **860**.

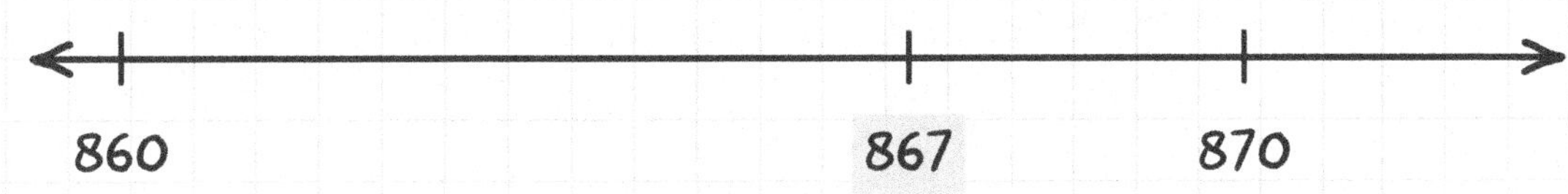

69. To solve, students should start at the number **19** on the number line and count on **7** spaces. They should land on **26**. The number line should look like this:

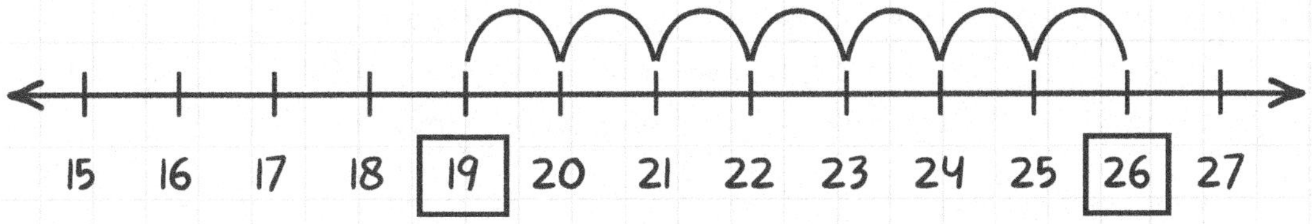

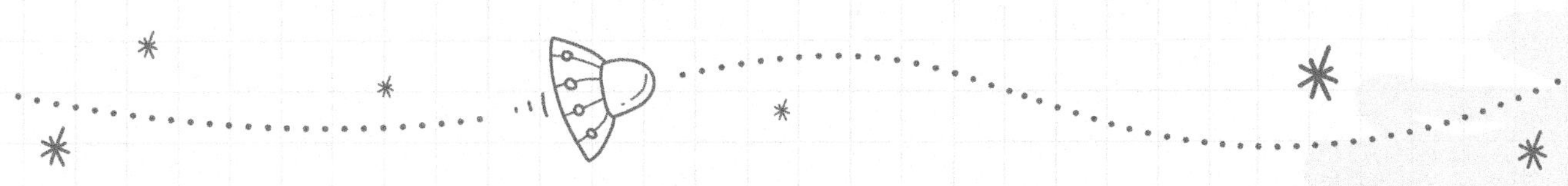

70. To solve, students should start at the number **274** on the number line and see how many spaces they need to count on to reach **286**. It should take **12** spaces to count to **286**. The number line should look like this:

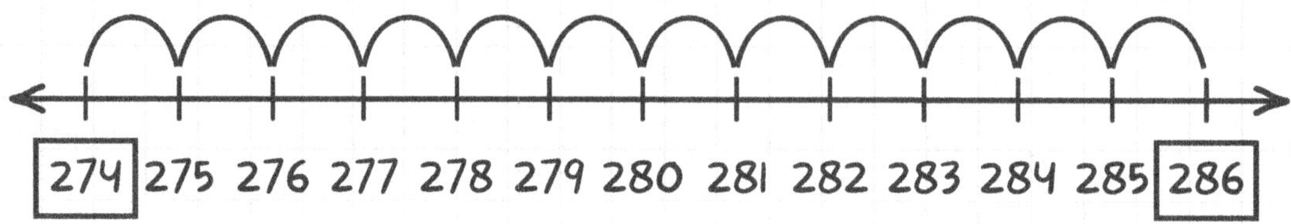

71. To solve, students should start at the number **621** on the number line and count back **8** spaces. They should land on **613**. The number line should look like this:

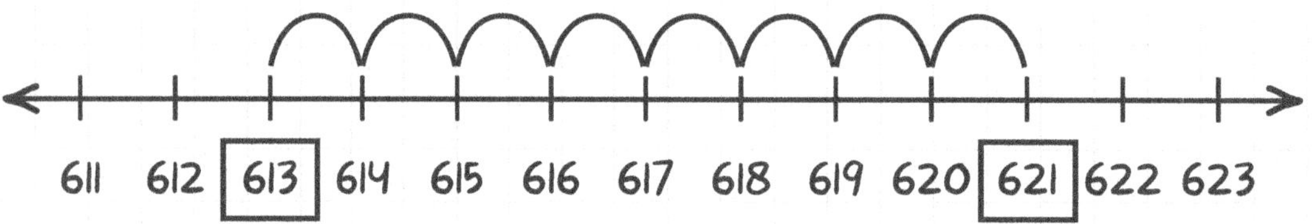

72. To solve, students should start at the number **501** on the number line and see how many spaces they need to count back to reach **490**. It should take **11** spaces to count back to **490**. The number line should look like this:

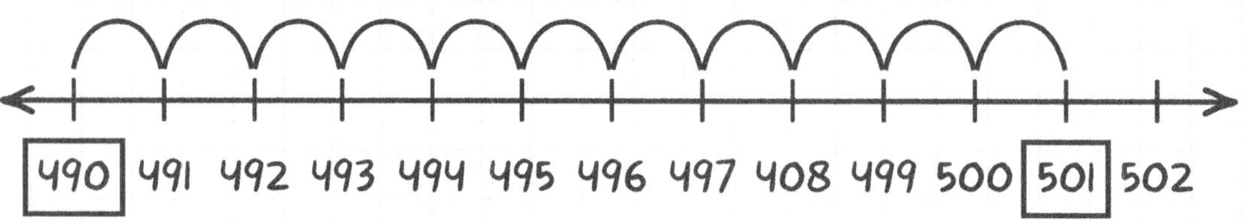

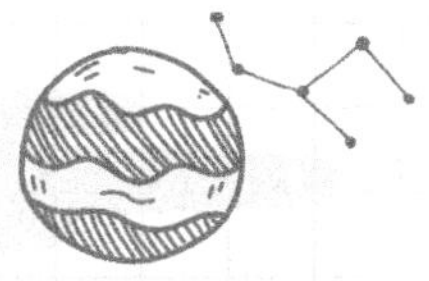

73. This number line starts at 44 and counts on 8 spaces to 52. The corresponding addition sentence would be 44 + 8 = 52.

74. This number line starts at 709 and counts on 4 spaces to 713. The corresponding addition sentence would be 709 + 4 = 713.

75. This number line starts at 366 and counts back 8 spaces to 358. The corresponding subtraction sentence would be 366 - 8 = 358.

76. This number line starts at 132 and counts back 7 spaces to 125. The corresponding subtraction sentence would be 132 - 7 = 125.

77. To solve, students should start at the number 62 on the number line and count back 25 spaces. They should land on 37. The number line should look like this:

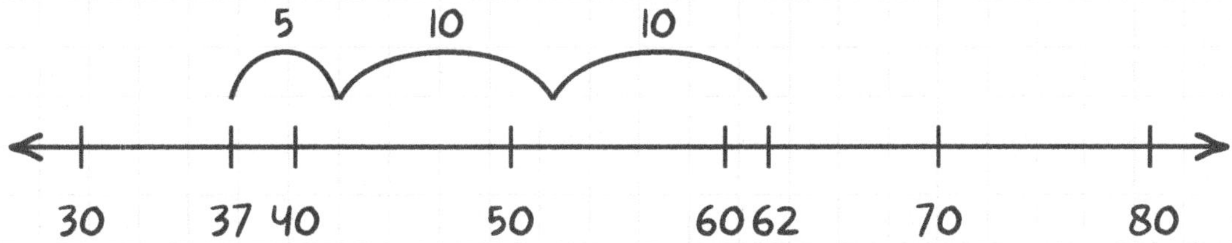

78. To solve, students should start at the number **325** on the number line and count back **58** spaces. They should land on **267**. The number line should look like this:

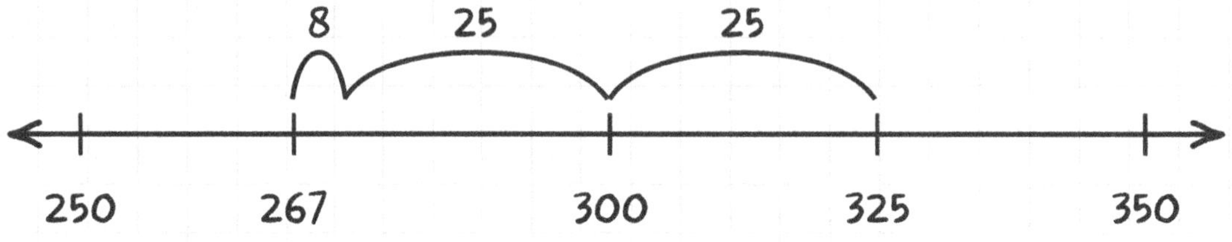

79. To solve, students should start at the number **86** on the number line and count on **102** spaces. They should land on **188**. The number line should look like this:

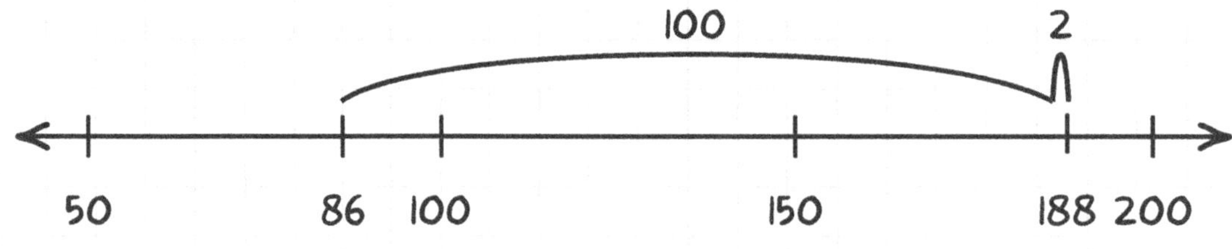

80. To solve, students should start at the number **612** on the number line and count on **99** spaces. They should land on **711**. Students could jump **100** spaces forward and then **1** back. The number line should look like this:

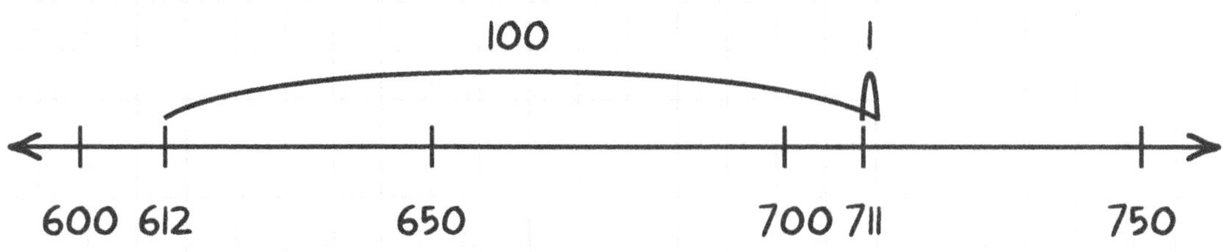

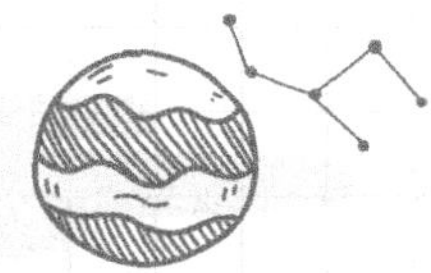

81. A.
The clock shows 4:00 because the hour hand is on the 4 and the minute hand is on the 12.

82. C.
The clock shows 7:10 because the hour hand is on the 7 and the minute hand is on the 2 (the 2 represents 10 minutes on a clock).

83. D.
The clock shows 8:45 because the hour hand is between the 8 and 9 and the minute hand is on the 9 (the 9 represents 45 minutes on a clock).

84. D.
The clock shows 12:30 because the hour hand is between the 12 and 1 and the minute hand is on the 6 (the 6 represents 30 minutes on a clock).

85. D.
The clock shows 10:20 because the hour hand is on the 10 and the minute hand is on the 4 (the 4 represents 20 minutes on a clock).

86.

The hour hand should be half way between the 2 and 3 and the minute hand should be on the 6.

87.

The hour hand should be on the 6 and the minute hand should be on the 3.

88.

The hour hand should be on the 11 and the minute hand should be on the 12.

89.

The hour hand should be half way between the 3 and 4 and the minute hand should be on the 6.

90.

The hour hand should be between the 4 and 5 and the minute hand should be on the 9.

91. B.
Half past is another way of denoting 30 minutes. Half past 4 is that same as 4:30.

92. C.
Quarter past is another way of denoting 15 minutes. Quarter past 5 is that same as 5:15.

ANSWER AND EXPLANATION

SECTION 3: TIME AND MONEY

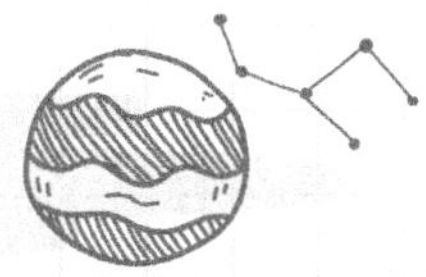

93. C.

At 12:55 there is 5 minutes left in the 12:00 hour. That means it is 5 minutes until 1:00.

94. A.

At 2:05 it is 5 minutes into the 2:00 hour. That means it is 5 minutes after 2:00.

95. C.

Quarter of is another way of denoting 45 minutes. Quarter of 2 is the same as 1:45.

96. D.

To solve this problem, students need to take the time at Carly woke up (7:00 am) and then add the 2 half hours together to get 1 whole hour. 7:00am+ 1 hour= 8:00am.

97. C.

If you counting in 15 minute increments, 12:15 would be the next train that is available.

98. D.

Students can start at 3:00 and count on 2 hours until they reach 5:00.

99. B.

Students should add an hour to the start time of class (4:15) and get 5:15. If you add on an additional half hour, you get 5:45. 5:45 is the same as quarter of 6.

100. A

Students should add the hours first 5 + 6 = 11 and then add the minutes 0 + 10 = 10 to determine that David will land in Florida at 11 : 10 pm.

101. B.
25¢ + 10¢ + 10¢ + 1¢ + 1¢ + 1¢ = 48¢

102. D.
5¢ + 5¢ + 5¢ + 10¢ + 1¢ + 1¢ + 1¢ + + 1¢ = 29¢

103. A.
25¢ + 25¢ + 10¢ + 10¢ + 5¢ + 5¢ + + 5¢ = 85¢

104. D
25¢ + 25¢ + 25¢ + 1¢ = 76¢

105. C.
25¢ + 25¢ + 10¢ + 10¢ + 10¢ + + 10¢ + 10¢ = 100¢

106. B.
Mary has 38¢. When she earned 50¢, you add 38¢ + 50¢ = 88¢.

107. C.
Anna has 95¢. When she gave away 25¢, you subtract 95¢ - 25¢ = 70¢.

108. D.
Matt has 92¢. When he gave away 50¢, you subtract 92¢ - 50¢ = 42¢.

109. A
Charlie has 33¢. When he found 25¢, you add 33¢ + 25¢ = 58¢.

110. D
Sam has 100¢. When he gave away 75¢, you subtract 100¢ - 75¢ = 25¢.

111. Donna has 80¢. 80¢ is more than 75¢, so she has enough money to buy the stickers.

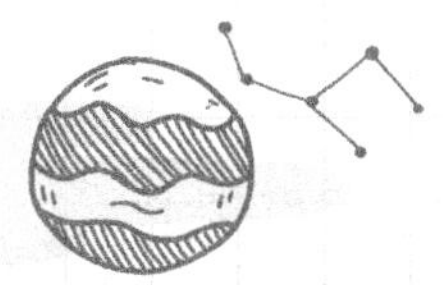

112. Jimmy has 100¢. 100¢ is more than 99¢, so he has enough money to buy the bell.

113. The cookie cost 50¢ and Liz gave the cashier 80¢. To solve, you need to subtract 80¢ - 50¢ = 30¢. Liz gets back 30¢ change.

114. Students may mentally know that 9 10s is 90. So you would need 9 dimes to make 90¢. Students may also draw a picture and count the money. A sample may look like:

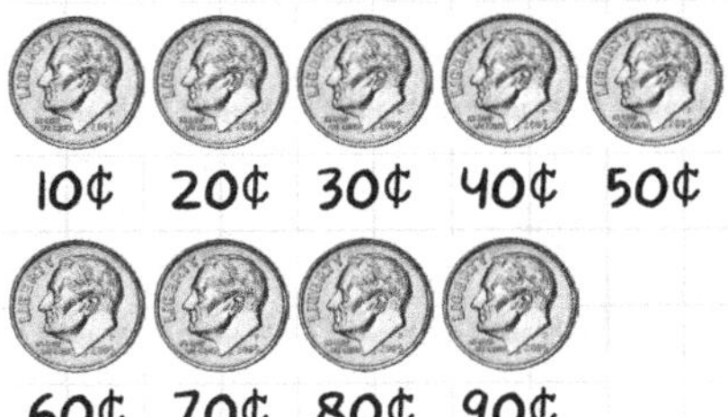

115. Students could write an addition sentence to solve 15¢ + 15¢ + 15¢ = = 45¢. Students may also draw a picture and count the money. A sample may look like:

116. To figure out the price of all 10 nails, Students may mentally know that 10 10s is 100. So Chris would need 100¢ to buy them. To solve, subtract 100 -85 =15 . Chris needs 15¢ more.

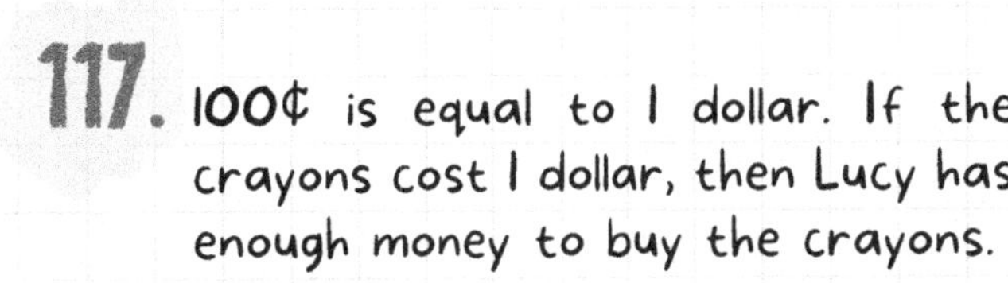

117. 100¢ is equal to 1 dollar. If the crayons cost 1 dollar, then Lucy has enough money to buy the crayons.

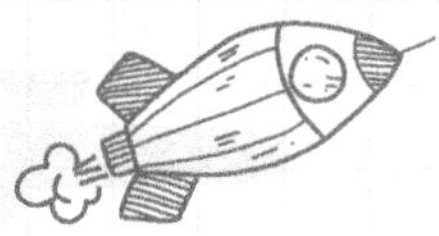

118. 7¢ more

To solve, students need to write a subtraction equation to compare the amounts of money. The subtraction equation should be: **68¢ - 61¢ = 7¢**. Parker has 7¢ more than Harry.

119. This is a two step equation. First, students have to figure out how much the items cost. This addition equation is **50¢ + 25¢ = 75¢**.

Step two is to figure out how much money Danny gave the cashier. This is a subtraction equation. **? - 75¢ = 25¢**. Students can use the related fact **75¢ + 25¢ = 100¢** (or **$1**). Danny gave the cashier **100¢** (or **$1**).

120. This is a two step equation. Step one is to figure out the total cost of Megan's lunch. Students will add together the cost of the items in the equation **50¢ + 25¢ + 10¢ = 85¢**.

Step two is to figure out how much money Megan has left to see if she can buy the **20¢** brownie. This is the subtraction equation **100¢ - 85¢ = 15¢**. Since **15¢** is less than **20¢**, Megan cannot buy the brownie.

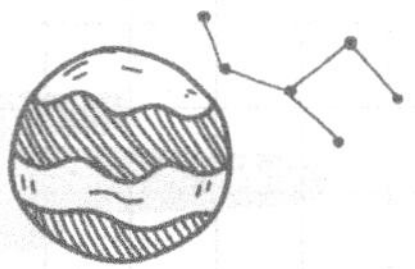

121. 60 minutes

Since there are more X's above 60 minutes than any other amount of time, 60 minutes is the most popular amount of time for most children to play outside.

122. 65 minutes

Since there are 0 X's above 65 minutes, 65 minutes is the least popular amount of time for most children to play outside.

123. 6 children

There are 6 X's above 45 minutes, so there are 6 children outside for 45 minutes a day.

124. 3 children

There are 3 X's above 20 minutes, so there are 3 children outside for 20 minutes a day.

125. 6 children

To solve, you need to count the X's above 50 and 55 minutes. There are 6 X's in all, so there are 6 children outside for 50 and 55 minutes.

126. 4 stickers

There are 4 X's above 50¢, so there were 4 50¢ stickers sold in a day.

127. 80¢

There are 6 X's above 5¢, so they made 30¢ total. There are 5 X's above 10¢, so they made 50¢ total. To solve, students need to combine the totals in an addition sentence: 30¢ + 50¢ = 80¢. 80¢ was made on 5¢ stickers and 10¢ stickers combined.

128. 3 more stickers

To solve, students need to write a subtraction equation to compare 5¢ stickers to 60¢ stickers. The subtraction equation should be: **6 - 3 = 3** stickers. **3** more **5¢** stickers were sold than **60¢** stickers.

129. 25¢ more

There are 0 X's above 35¢, so they made 0¢ total. There is 1 X above 25¢, so they made 25¢ total. To solve, students need to write a subtraction equation to compare the amount of money 35¢ stickers made to the amount of money 25¢ stickers made. The subtraction equation should be: **25¢ - 0¢ = 25¢**. They made **25¢** more on **25¢** stickers compared to **35¢** stickers.

130. 4 more stickers

To solve, students need to write a subtraction equation to compare 65¢ stickers to 10¢ stickers. The subtraction equation should be: **5 - 1 = 4** stickers. **4** more **10¢** stickers were sold than **65¢** stickers.

The line plot should look like this:

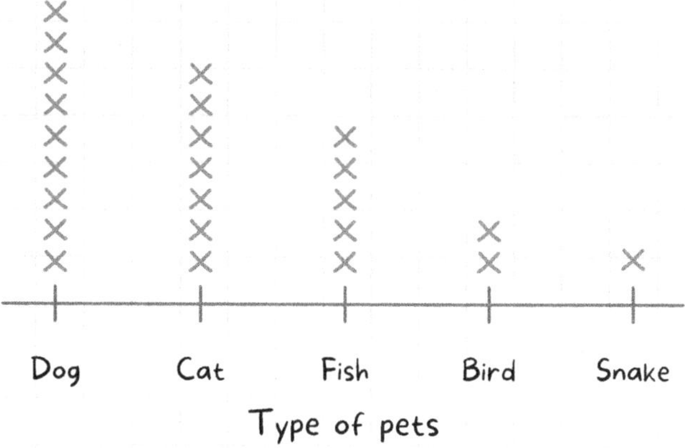

Dog	9
Cat	7
Fish	5
Bird	2
Snake	1

131. Dogs

The most X's are above dog, so most amount of children have dogs as pets.

ANSWER AND EXPLANATION

SECTION 4: REPRESENTING AND INTERPRETING DATA

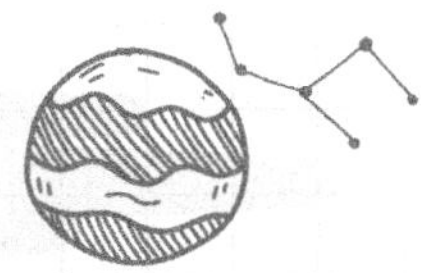

132. Snakes

The least X's are above snake, so the least amount of children have snakes as pets.

133. 16 dogs and cats combined

There are 9 X's above dogs and there are 7 X's above cats. To solve, students need to combine the totals in an addition sentence: 9 + 7 = 16. There are 16 children who have dogs and cats combined.

134. 5 more cats

To solve, students need to write a subtraction equation to compare cats to birds. The subtraction equation should be: 7 - 2 = 5. There are 5 more cats as pets than birds.

135. 8 fewer snakes

To solve, students need to write a subtraction equation to compare snakes to dogs. The subtraction equation should be: 9 - 1 = 8. There are 8 fewer snakes than dogs as pets.

The line plot should look like this:

How tall are you?

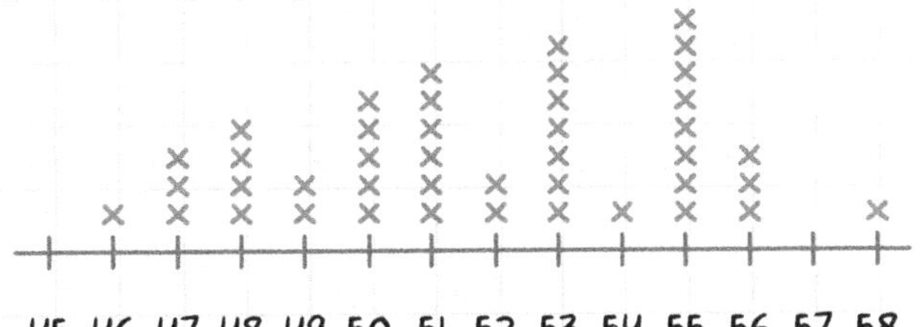

Height in inches

136. 7 kids

There are 7 X's above 53 inches, so there are 7 kids that are 53 inches tall.

137. 8 kids

There are 8 X's above 55 inches, so there are 8 kids that are 55 inches tall.

138. 15 kids
There are 7 X's above 53 inches and there are 8 X's above 55 inches. To solve, students need to combine the totals in an addition sentence: 7 + 8 = 15. There are 15 kids that are 53 inches and 54 inches combined.

139. 15 kids
To solve, students need to add all of the X's that are below 51 inches. The addition sentence is 1 + 3 + 4 + 2 + 5 = 15 kids. There are 15 kids shorter than 51 inches.

140. 13 kids
To solve, students need to add all of the X's that are above 53 inches. The addition sentence is 1 + 8 + 3 + 1 = 13 kids. There are 13 kids taller than 53 inches.

141. 40 more points
In the first quarter, Team A scored 20 points, and Team B scored 60 points. To solve, students need to subtract the points to compare. The subtraction equation is 60 - 20 = 40 points. Team B scored 40 more points than Team A.

142. 14 fewer points
In the fourth quarter, Team A scored about 38 points, and Team B scored about 62 points. To solve, students need to subtract the points to compare. The subtraction equation is 62 - 38 = 24 points. Team A scored 24 fewer points than Team B.

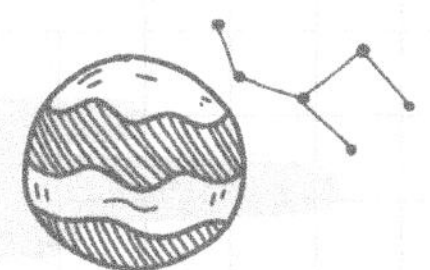

143. 78 points combined

To solve, students will add the points from the third and fourth quarters for Team A. The addition sentence is 40 + 38 = 78 points. Team A scored 78 points in the third and fourth quarters combined.

144. 45 points combined

To solve, students will add the points from Team A and Team B in the second quarter. The addition sentence is 30 + 15 = 45 points. Teams A and B scored about 45 points combined in the second quarter.

145. Team B won with 185 points

To figure out the final score, students need to add all the points for each team. For Team A, the addition equation is 20 + 30 + 40 + 38 = 128 points. For Team B, the addition equation is 60 + 15 + 48 + 62 = 185 point. Since 185 is greater than 128, Team B won the game.

146. Soccer

Since there are more soccer balls than any other picture, soccer is the sport that most kids like best.

147. Tennis

Since there are fewer tennis rackets than any other picture, tennis is the sport that the least amount of kids like.

148. 12 kids

To solve, students will add the amount of kids who like soccer and the amount of kids who like baseball. The addition sentence should be: 5 + 7 = 12 kids. There are 12 kids who like soccer and baseball combined.

149. 4 kids

2 kids picked tennis. To solve, students need to add 2 additional kids. The addition equation is 2 + 2 = 4. If 2 more kids picked tennis then 4 kids would have picked tennis in all.

150. 3 fewer kids

Students need to subtract to compare the amount of kids who liked hockey and soccer. The subtraction equation is 7 - 4 = 3 kids. 3 fewer kids like hockey compared to soccer.

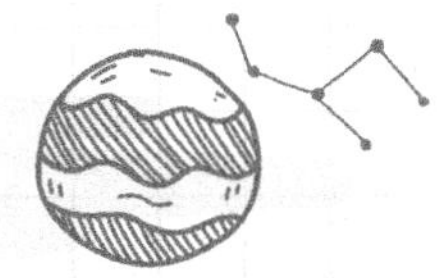

The picture graph should look like this:

What did you have for lunch?

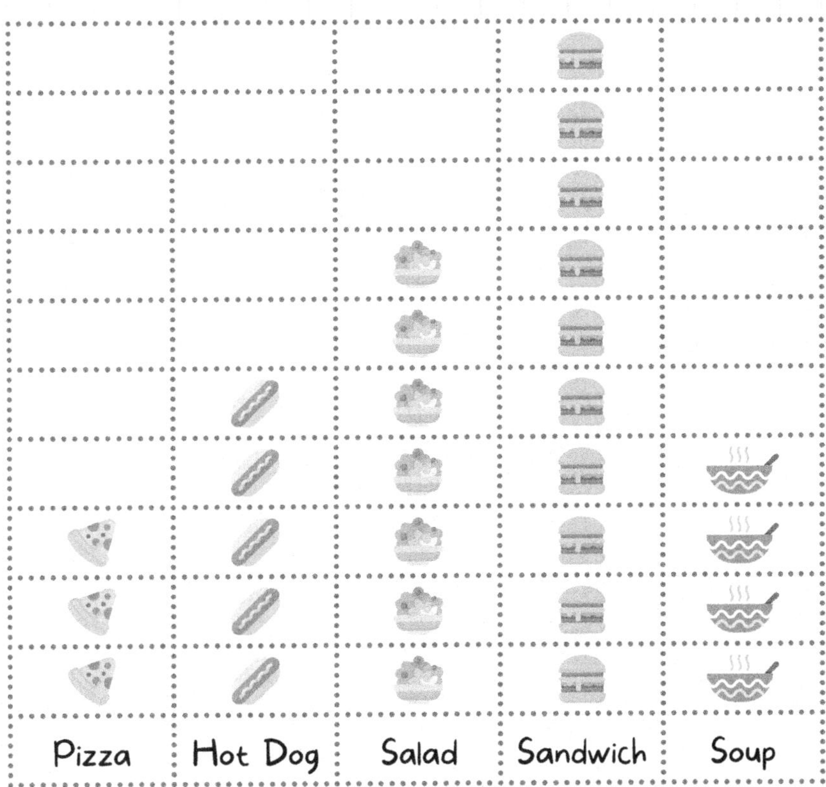

151. 6 fewer people
Students need to subtract to compare the amount of people who had soup instead of sandwich for lunch. The subtraction equation is 10 - 4 = 6 people. 6 fewer people had soup compared to a sandwich for lunch.

152. 17 people
To solve, students will add the number of people who had a sandwich and the amount of people who had a salad. The addition sentence is 10 + 7 = 17 people. There are 17 people who had a sandwich and a salad combined.

153. Sandwich

Since there are more sandwiches than any other picture, sandwich is the lunch that most people had.

154. Pizza

Since there are fewer pizzas than any other picture, pizza is the lunch that the least amount of people chose.

155. 7 more pizzas

3 people picked pizza, and 10 people picked sandwich. To solve, students can start at 3 and count to on to 10 to see how many more people would have to pick pizza to make it equal to sandwich. The answer is 7. Students can also write the subtraction sentence: 10 - 3 = 7 pizzas. If there were 7 more pizzas, then pizza and sandwich would be equal.

The bar graph should look like this:

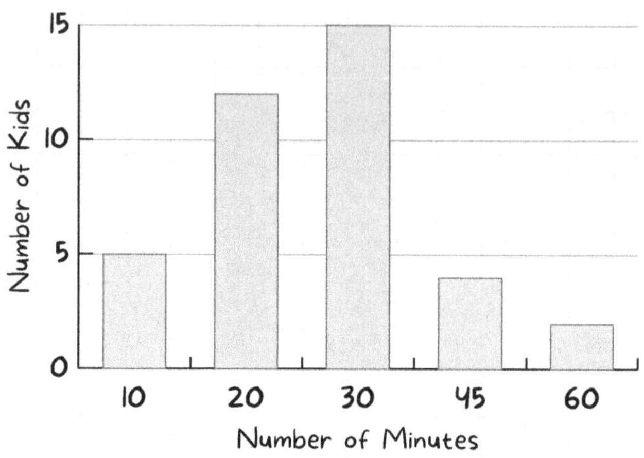

156. 4 kids

Since the bar above 45 minutes does not quite reach 5, there are approximately 4 kids who read for 45 minutes a day.

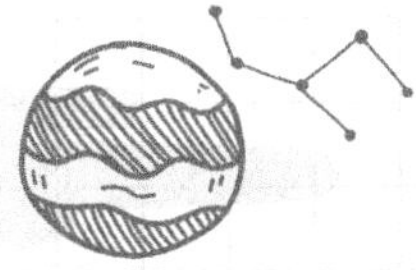

157. 3 more kids

Students need to subtract to compare the amount of kids who read for 30 minutes and 20 minutes. The subtraction equation is 15 - 12 = 3 kids. 3 more kids read for 30 minutes than 20 minutes.

158. 30 minutes

Since the bar for 30 minutes is the highest, 30 minutes is the most popular amount of time for kids to read.

159. 17 kids

To solve, students need to add together all the kids who read for less than 30 minutes a day. The addition equation is 12 + 5 = 17 kids. There are 17 kids who read for less than 30 minutes a day.

160. 21 kids

To solve, students need to add together all the kids who read for more than 20 minutes a day. The addition equation is 15 + 4 + 2 = 21 kids. There are 21 kids who read for more than 20 minutes a day.

ANSWER AND EXPLANATION

SECTION 1: SHAPES AND THEIR ATTRIBUTES

1. Polygons are closed, flat shapes with straight sides.

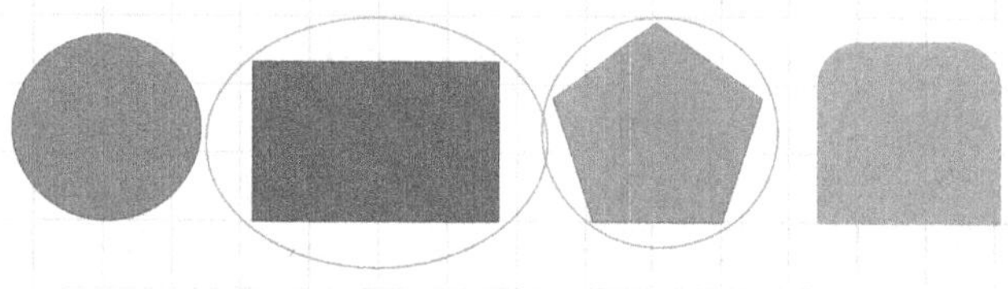

2. Polygons are closed, flat shapes with straight sides.

3. Polygons are closed, flat shapes with straight sides.

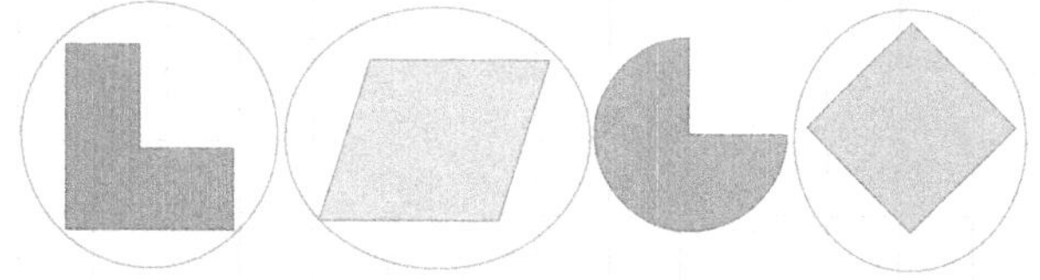

4. Polygons are closed, flat shapes with straight sides.

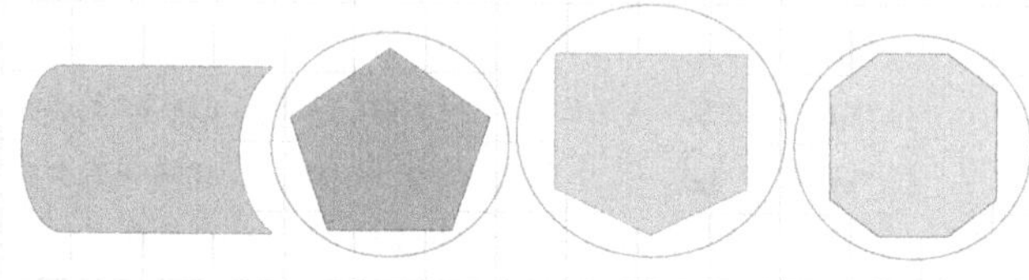

5. A triangle has 3 straight sides and 3 vertices.

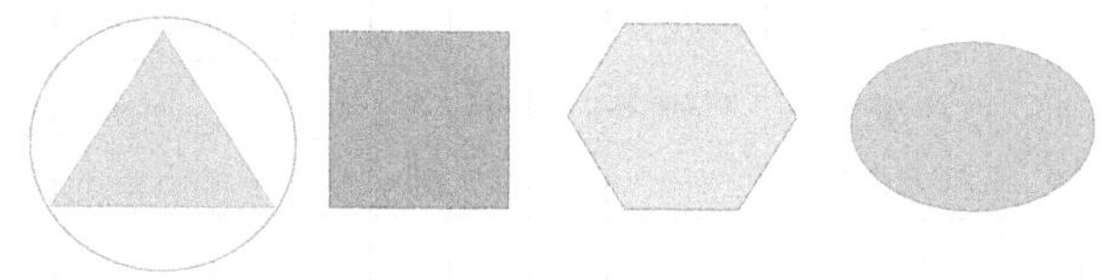

6. A square has 4 sides and 4 vertices.

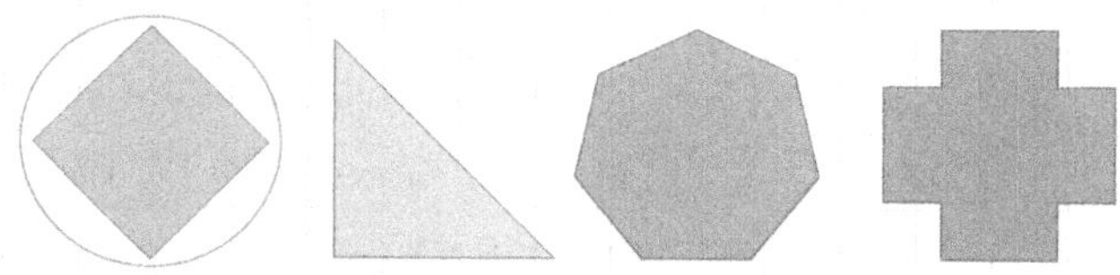

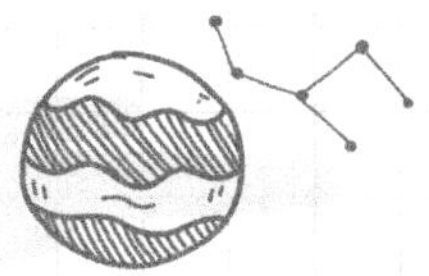

7. An octagon has 8 sides and 8 vertices.

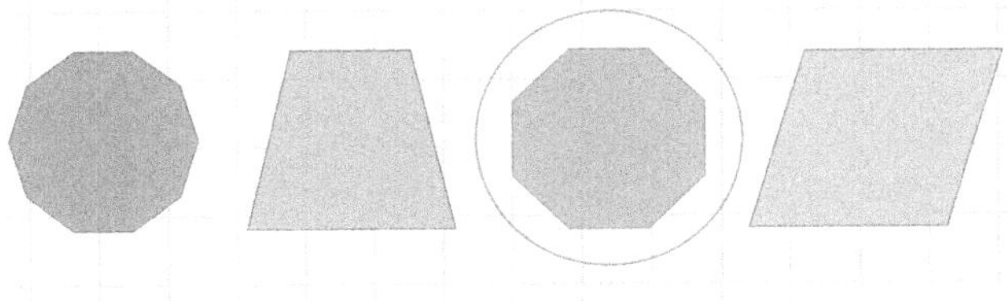

8. A pentagon has 5 sides and 5 vertices.

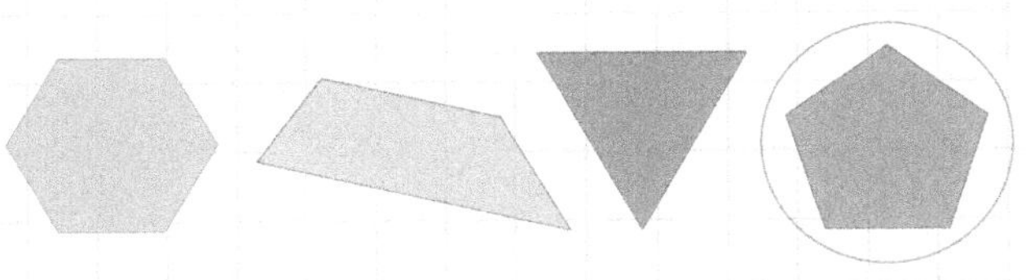

9. Students should draw a non-polygon shape that is open or doesn't have straight sides. An example may be:

10. Students should draw any polygon with 4 sides and 4 vertices. An example may be:

11. Students should draw a non-polygon shape with 5 sides that is open or doesn't have straight sides. An example may be:

12. Students should draw any polygon with 5 sides and 5 vertices. An example may be:

13. Students should draw an open, non-polygon shape with a curve. An example may be:

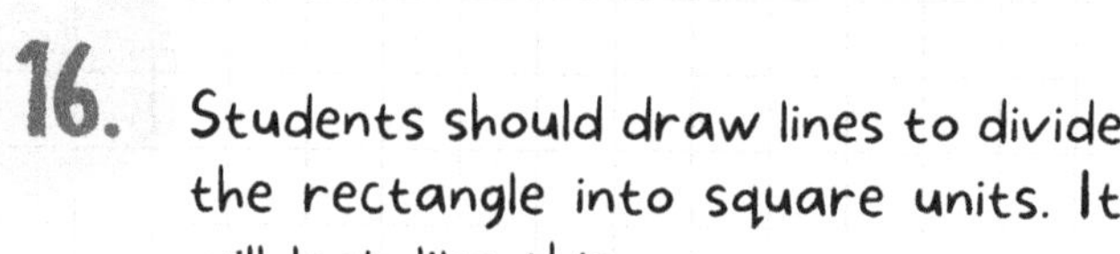

14. Students should draw lines to divide the rectangle into square units. It will look like this:

15. There are 40 square units in the rectangle above.

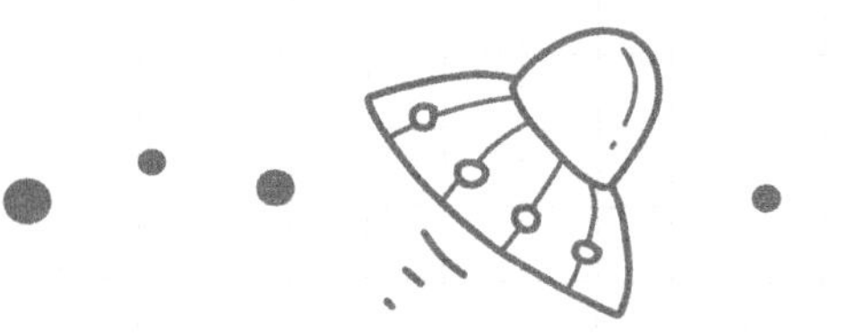

16. Students should draw lines to divide the rectangle into square units. It will look like this:

17. There are 6 square units in the rectangle above.

18. Students should draw lines to divide the rectangle into square units. It will look like this:

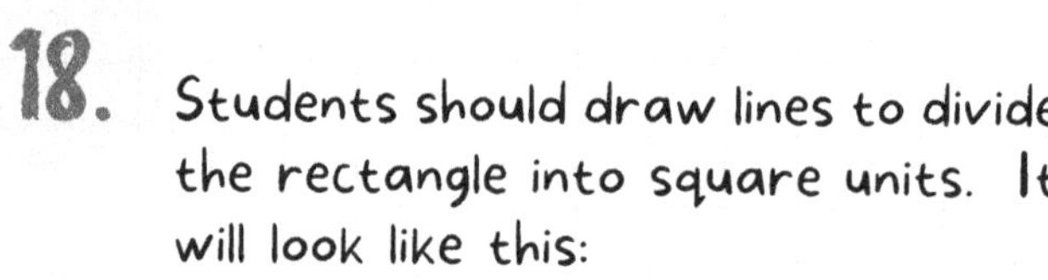

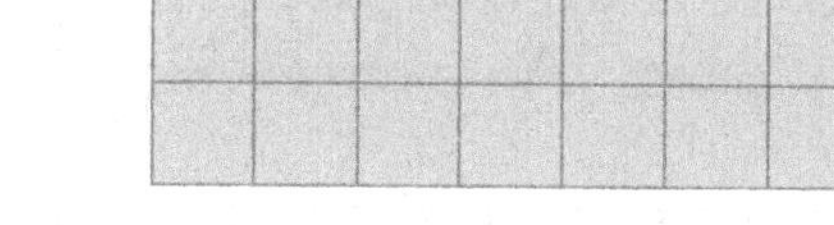

19. There are 18 square units in the rectangle above.

ANSWER AND EXPLANATION

SECTION 1: SHAPES AND THEIR ATTRIBUTES

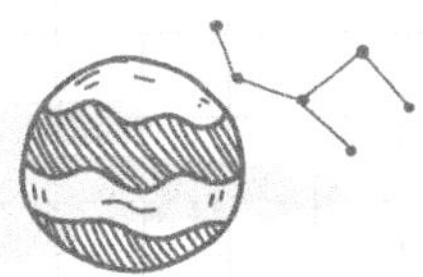

20. Columns are division that go from top to bottom. Students should draw 4 equal columns that look like this:

21. Rows are divisions that go from side to side. Students should show 5 equal rows that look like this:

22. Columns are division that go from top to bottom. Students should draw 3 equal columns that look like this:

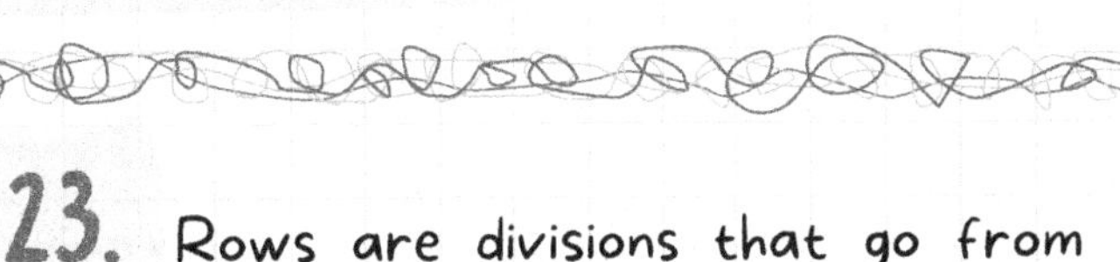

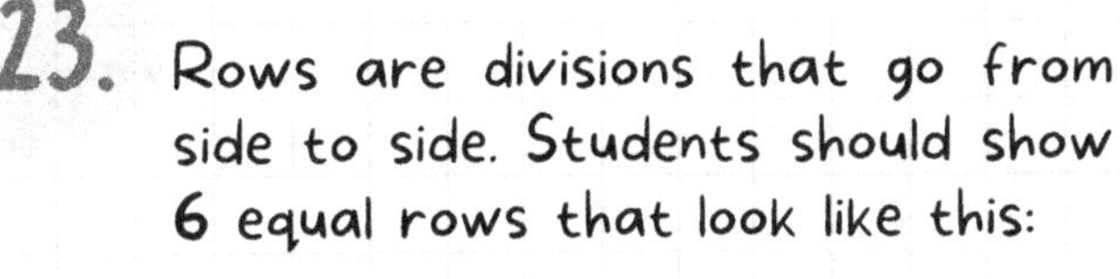

23. Rows are divisions that go from side to side. Students should show 6 equal rows that look like this:

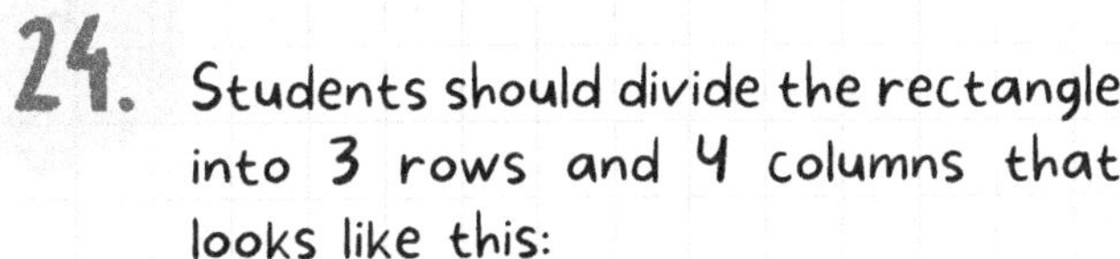

24. Students should divide the rectangle into 3 rows and 4 columns that looks like this:

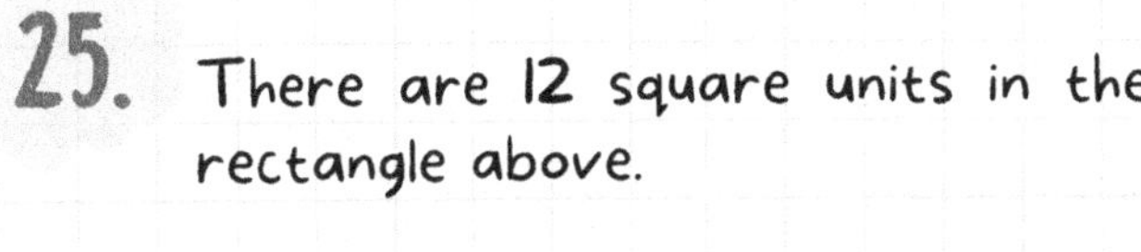

25. There are 12 square units in the rectangle above.

26. Students should divide the rectangle into 2 rows and 5 columns that looks like this:

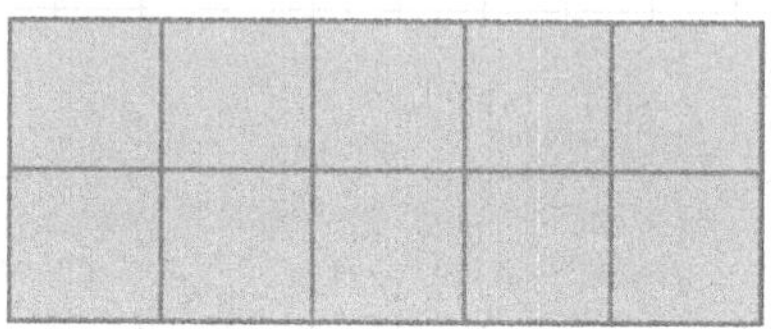

27. There are 10 square units in the rectangle above.

28. In half means 2 equal parts. Students can divide the circles in half in many different ways. An example may be:

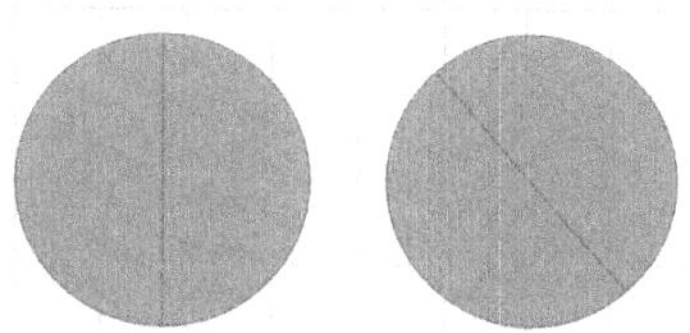

29. In half means 2 equal parts. Students can divide the rectangles in half in many different ways. It is important to note that all parts must be equal. An example may be:

30. In fourths means 4 equal parts. Students can divide the rectangles in fourths in many different ways. It is important to note that all parts must be equal. An example may be:

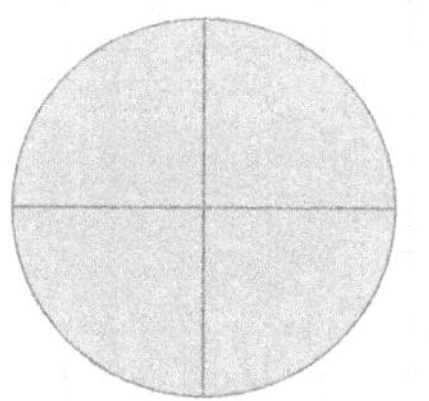

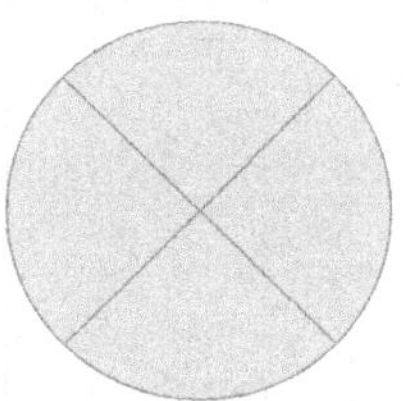

ANSWER AND EXPLANATION

SECTION 1: SHAPES AND THEIR ATTRIBUTES

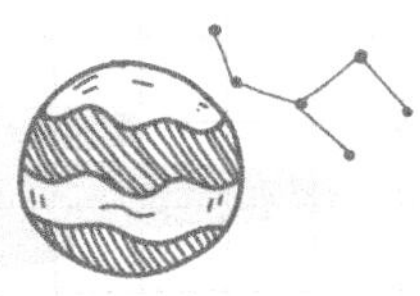

31. In fourths means 4 equal parts. Students can divide the rectangles in fourths in many different ways. It is important to note that all parts must be equal. An example may be:

32. In fourths means 4 equal parts. The only shape divided into 4 equal parts is the circle.

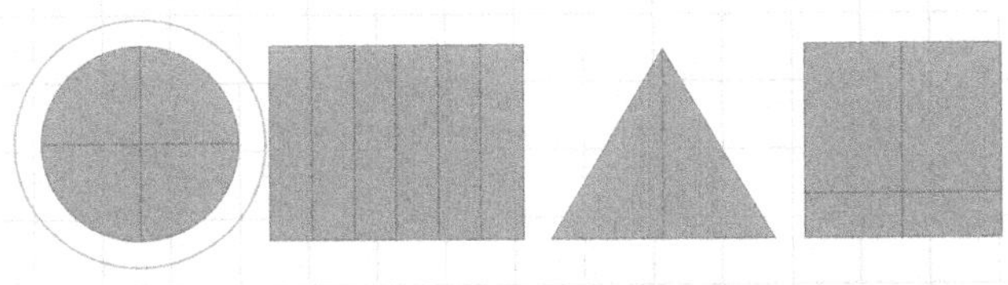

33. In half means 2 equal parts. The only shape divided into 2 equal parts is the hexagon.

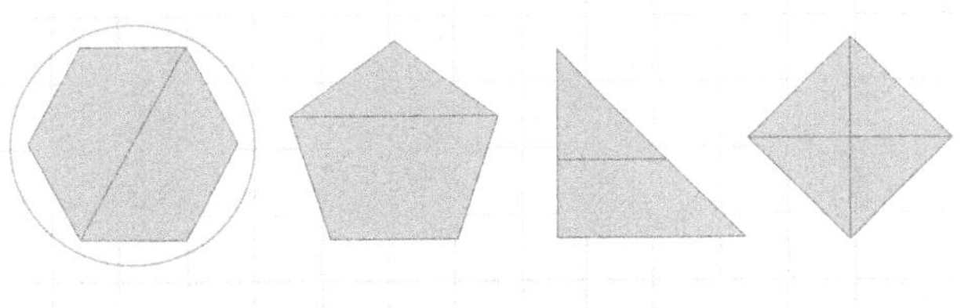

34. In thirds means 3 equal parts. The only shape divided into 3 equal parts is the rectangle.

35. A.
The square is divided into 4 equal parts or fourths.

36. B.
The trapezoid is divided into 2 equal parts or halves.

37. D.
The square is divided into 3 equal parts or thirds.

38.

39. 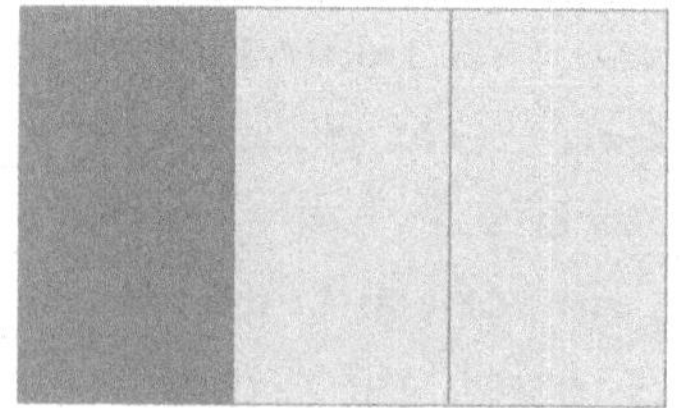

40. 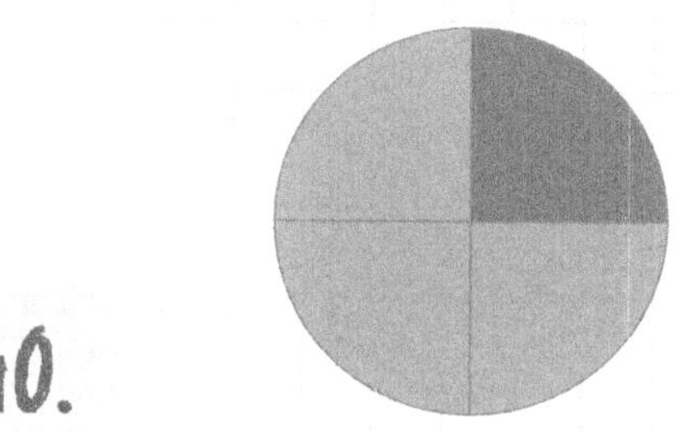

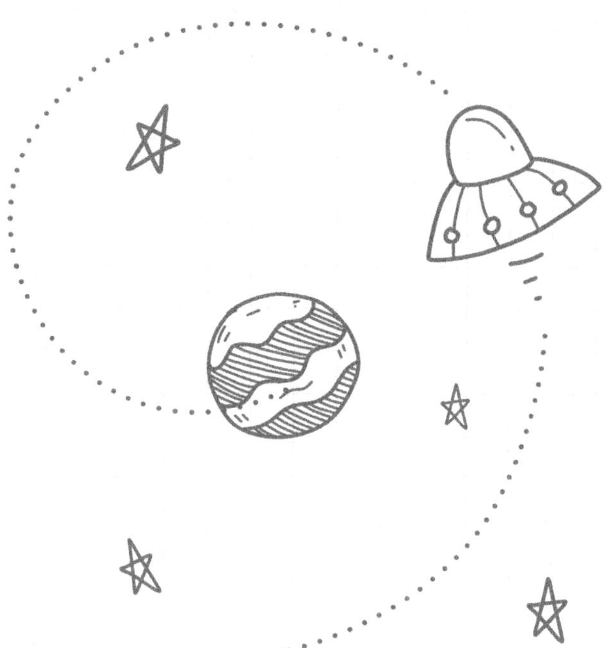

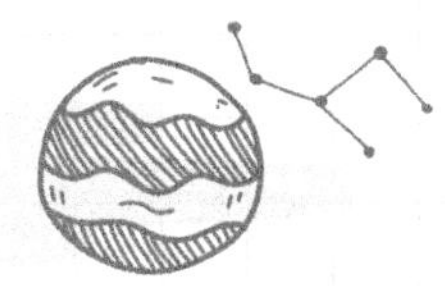

1. C.
The first group has 10, and the second group has 8. When you join them together and count them in all, the sum is 18. The corresponding addition equation is 10 + 8 = 18

2. Any visual representation that depicts a strategy to show that the missing number is 54 is correct. Students can visually show 82 - 28 = 54 or students can draw 28 objects and count on 54 more until they reach 82. Students may draw representation in base 10 rods and units.

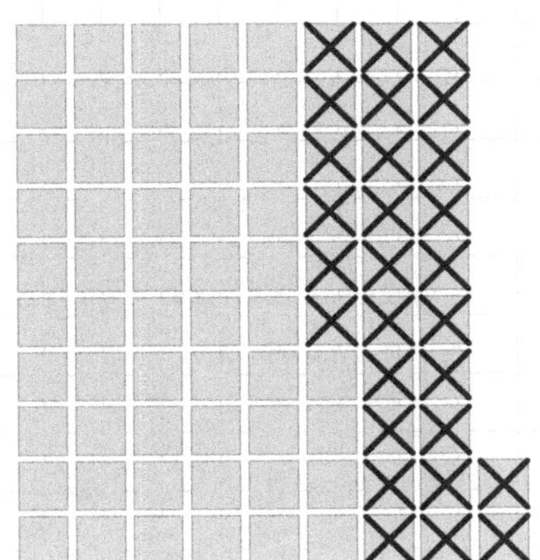

82 - 28 = 54

3. 69 acorns
This is a two step problem. First students will add the two groups of acorns. There are 45 in the first group and 37 in the second group. The equation is 45 + 37. To solve, students will add the ones: 7 + 5 = 12 and add the tens: 30 + 40 = 70. The sum is 82. Step two is to figure out the difference. The subtraction equation should be 82 - 13. The difference is 69. 82 - 13 = 69 acorns.

4. The group of smileys has 32 total. There are 28 smileys that are crossed out. There are 4 smileys remaining. The corresponding subtraction equation would be 32 - 28 = 4.

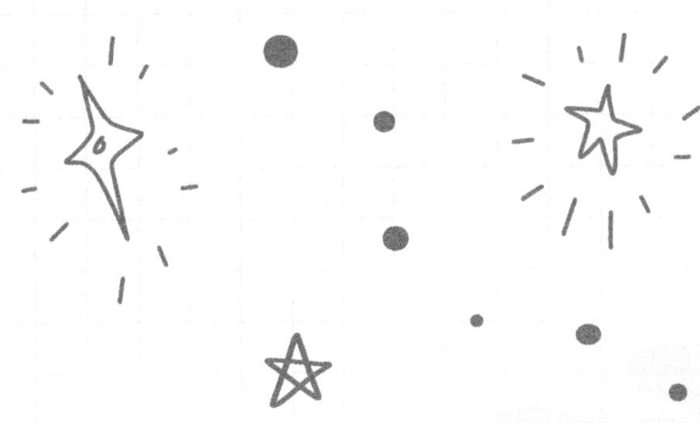

5. Any visual representation that shows that the answer is 5 is correct. Students can subtract 64 - 59 = 5, or students can draw a picture to compare and find the difference.

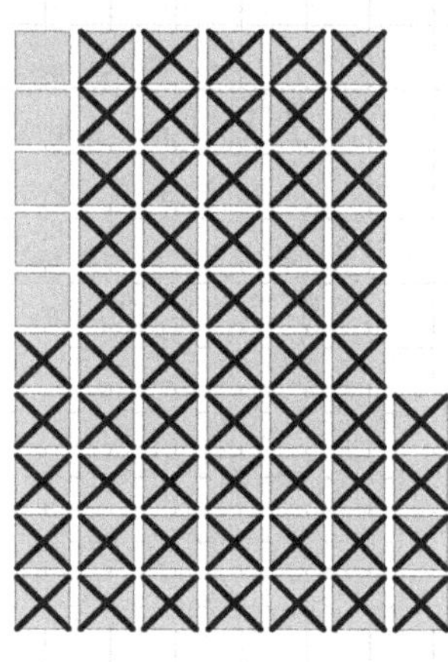

64 - 59 = 5

6. D.

8 + 9 = 17 is a near double addition equation. This means, having a double fact that is close to this memorized, would help you solve it quickly. If you knew that (D) 9 + 9 = 18, then you would know that subtracting 1 more would change the problem to 8 + 9 = 17.

7. A.

If you take one away from the first addend (10) and add it to the second addend (2), then you change 10 + 2 to 9 + 3. Both equations have the same sum of 12.

8. 8. The student can either write 15 - 1 = 14 or 15 - 14 = 1.

9.

9 + 10 = 19	13 + 5 = 18	17 - 12 = 5
4 + 12 = 16	6 + 6 = 12	15 - 6 = 9
19 - 8 = 11	12 - 10 = 2	1 + 14 = 15

10. Solve these addition equations with a missing addend.

12 + 7 = 19	6 + 11 = 17	2 + 15 = 17
8 + 6 = 14	5 + 5 = 10	6 + 9 = 15

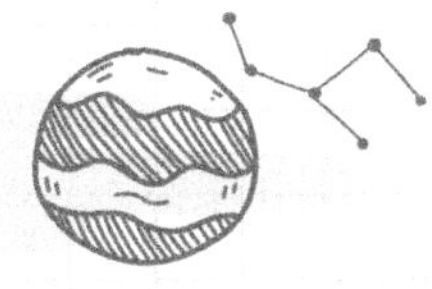

11. There are 11 watermelons. 11 is an odd number because all the watermelons do not have a partner.

12. Students should write the equation 4 + 4 = 8.

13. Any visual representation that shows 12 feet. For example:

14. 4 + 4 + 4 = 12

15. Any visual representation with 3 objects in 5 rows is correct. For example:

16. There are 4 tens in the number 741.

17. B.
The number 539 has 5 hundreds, 2 tens, and 9 ones.

18. 245, 250, 255, 260, 265, 270, 275, 280

19. The number 126 has 1 hundred, 2 tens, and 6 ones. When written in expanded form it is 126 = 100 + + 20 + 6.

20. 754 > 544. When comparing the hundreds, 754 has 7 hundreds and 544 only has 5 hundreds.

21.

62 + 35 = 97	27 + 35 = 62	55 + 40 = 95
38 + 24 = 62	64 + 26 = 90	39 + 17 = 56

22.

32 - 17 = 15	86 - 38 = 48	78 - 25 = 53
91 - 55 = 36	46 - 25 = 21	67 - 43 = 24

23. 17 + 21 + 5 + 46 = 89

To show their work, students can draw a picture using ten rods and units, they can show how they added the tens and ones together, or students can visually depict their addition strategy.

An example may be:

Adding the ones 7 + 1 + 5 + 6 = 19

Adding the tens 10 + 20 + 0 + 40 = = 70

Adding the tens and ones: 70 + 19 = = 89

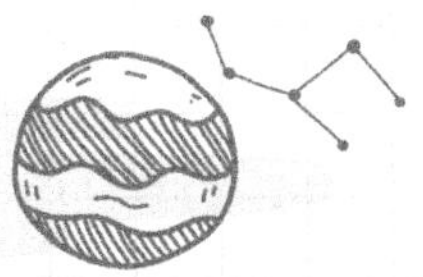

24. Students should write the addition equation 33 + 100 = 133 or 100 + 33 = 133.

25. 49 people
To solve the problem, students need to write a subtraction equation. Start with the total number of people who get free books (125) and subtract the people who were before Donald in line (76) to find out how many people were after him. The subtraction equation is 125 - 76 = 49 people. There are 49 people in line after Donald that will also get a free book.

26. C.
A measuring tape would be the most efficient tool because it is long enough to measure the height of the refrigerator. A measuring tape measures in feet and inches, which are the standard measurements closest to the height of a fridge.

27. From head to end, the caterpillar is approximately 4 centimeters long.

28. 6 fewer paperclips
To solve the problem, students need to write a subtraction equation to compare the inches to paperclips. The subtraction equation is 24 - 18 = 6 paper clips. It takes 6 fewer paperclips than inches to measure the water fountain.

29. D.
Since it took fewer yards to measure Christie's bathroom than feet, that means yards are bigger than feet.

30. C.

20 feet is the only answer choice that makes sense in regards to the height of a flagpole. The rest of the options would be too small or too big.

31. 48 inches taller

To solve the problem, students need to write a subtraction equation to compare the heights of the two different animals. The subtraction equation is 180 - 132 = 48 inches. A giraffe is 48 inches taller than an elephant.

32. Students should draw a line that looks like this:

3 inches | 3 inches

3 + 3 = 6 inches

33. Students should draw a picture that looks like this:

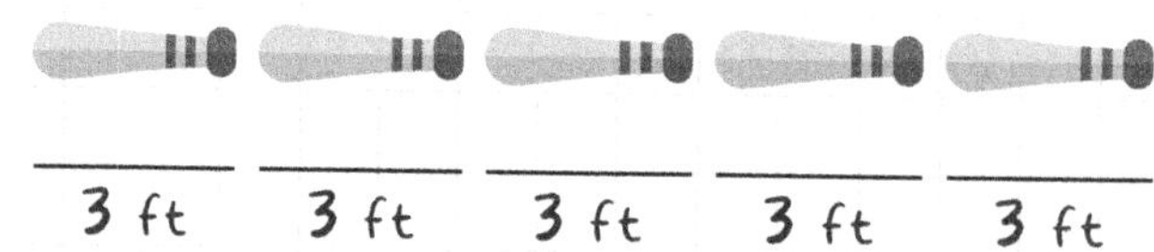

3 ft | 3 ft | 3 ft | 3 ft | 3 ft

Students can write an addition equation and solve: 3 + 3 + 3 + 3 + 3 = 15 feet.

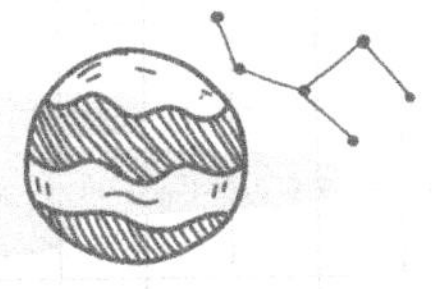

34. The number 267 should be placed closer to 275 than to 250.

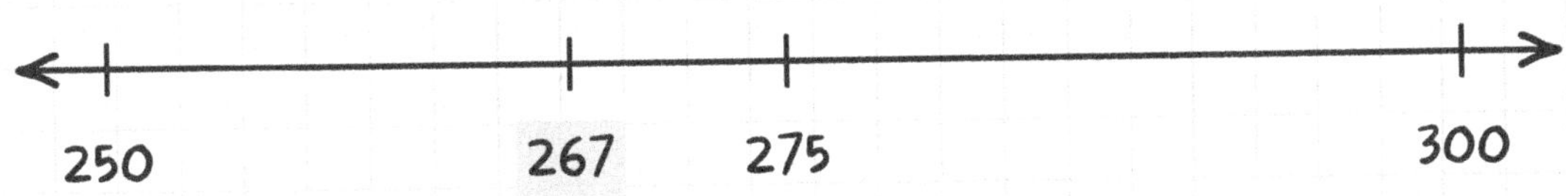

35. To solve, students should start at the number 521 on the number line and count back 10 spaces. They should land on 511. The number line should look like this:

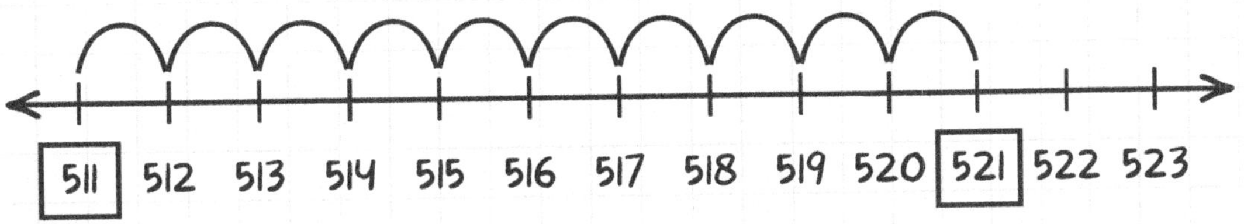

36. B.
The clock shows 10:25 because the hour hand is on the 10 and the minute hand is on the 5.

37.

The hour hand should be halfway between the 4 and 5, and the minute hand should be on the 9.

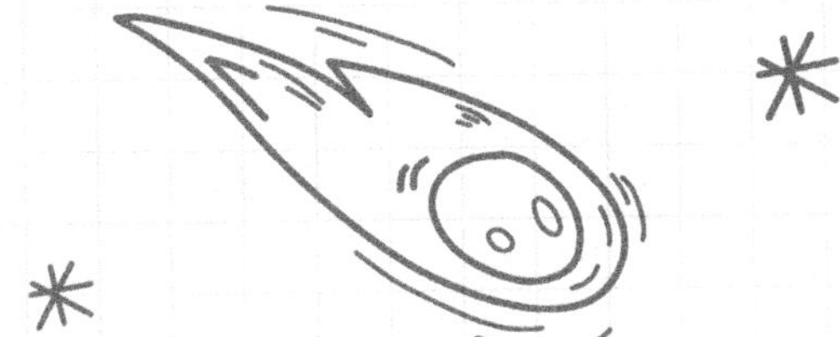

38. A.

Quarter past is another way of denoting 15 minutes. Quarter past 1 is that same as 1:15.

39. D.

To solve this problem, students need to take the time Kenny started packing (12:00 pm) and count on until be stopped at 5:00pm. 12:00 + 5 hours = 5:00 pm. It took Kenny 5 hours to pack.

40. B.

Half past of is another way of denoting 30 minutes. Half past 12 is the same as 12:30

41. C.

25¢ + 25¢ + 10¢ + 5¢+ 1¢ + 1¢+ 1¢ + + 1¢ + 1¢ = 70¢

42. C.

25¢ + 10¢ + 10¢ +10¢ + 5¢ + 1¢ = 61¢

43. A.

Erin has 78¢. To show how she gave 50¢ back to her brother, subtract 78¢ + 50¢ = 28¢.

44. 15¢ back

The apple and water combined cost 85¢ (35¢ + 50¢ = 85¢), and Tony gave the cashier 4 quarters which is equal to 100¢. To solve, you need to subtract 100¢ - 85¢ = = 15¢. Tony gets back 15¢ change.

45. 4 pencils

Students could write an addition sentence to solve 20¢ + 20¢ + + 20¢ + 20¢ = 80¢. Students may also draw a picture and count the money. A sample may look like:

20¢ 40¢

60¢ 80¢

Kelly can buy 4 pencils with her 80¢.

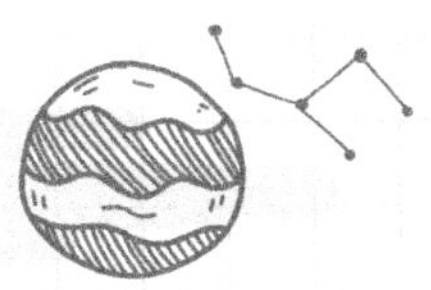

46. Polygons are closed, flat shapes with straight sides.

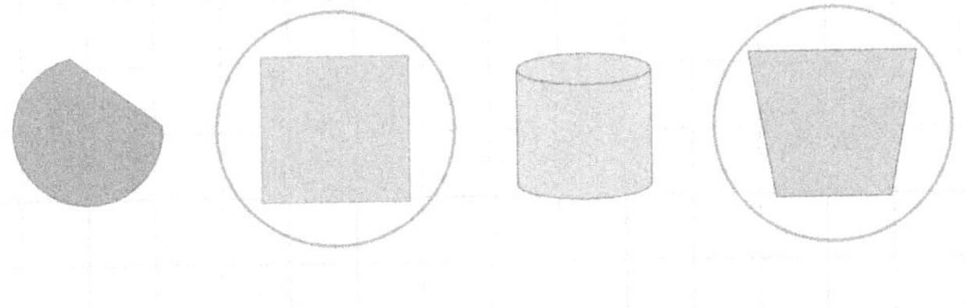

47. A hexagon has 6 straight sides and 6 vertices.

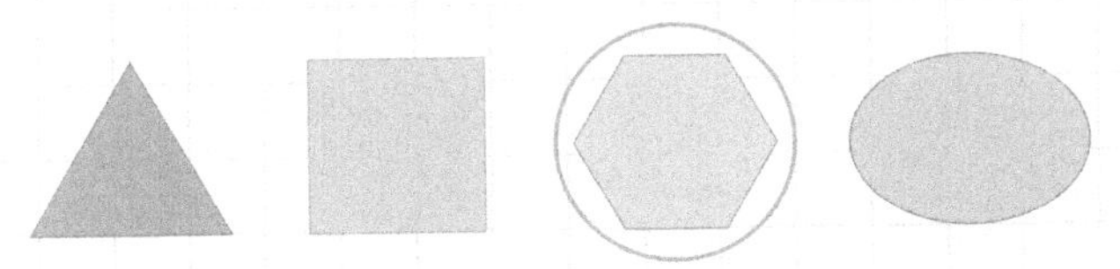

48. Students should draw a non-polygon shape with 3 sides that is open or doesn't have straight sides. An example may be:

49. Students should divide the square into 4 rows and 4 columns that looks like this:

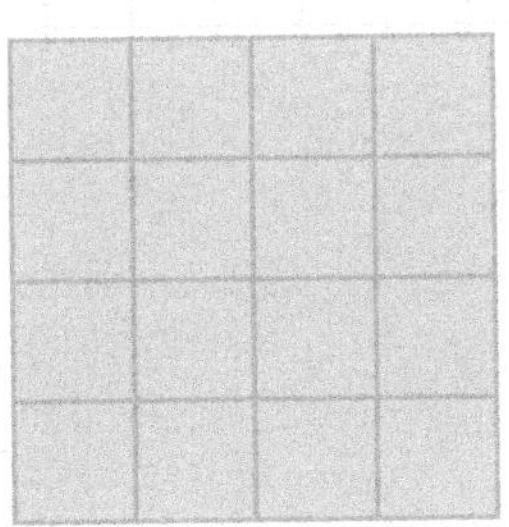

50. There are 16 square units in the rectangle above.

Made in the USA
Columbia, SC
24 March 2020

89899076R00141